Roberto Lanza · Antonio Meloni

The Earth's Magnetism
An Introduction for Geologists

Roberto Lanza · Antonio Meloni

The Earth's Magnetism

An Introduction for Geologists

With 167 Figures and 6 Tables

Authors

Prof. Dr. Roberto Lanza

Dipartimento di Scienze della Terra
Università di Torino
Via Valperga Caluso 35
10125 Torino, Italy

Phone: +39 011 6705165
Fax: +39 011 6705146
E-mail: roberto.lanza@unito.it

Dr. Antonio Meloni

Istituto Nazionale di Geofisica e Vulcanologia
Via di Vigna Murata 605
00143 Roma, Italy

Phone: +39 06 51860317
Fax: +39 06 51860397
E-mail: meloni@ingv.it

ISBN 978-3-642-06624-5   e-ISBN 978-3-540-27980-8

Springer is a part of Springer Science+Business Media
springeronline.com
© Springer-Verlag Berlin Heidelberg 2010
Printed in Germany

Cover design: Erich Kirchner, Heidelberg

Preface

Geomagnetism has always been at the forefront among the various branches of geophysics. At the end of the 16th century William Gilbert determined that the Earth is a big magnet, implying that it has a magnetic field; in the 1830s Carl Friedrich Gauss was able to formulate a procedure to measure the field completely and analyzed its characteristics with the spherical harmonic analysis, a method still used in the era of satellites and computers. Nevertheless, as recently as in the sixties, geophysics textbooks devoted only a thin chapter to geomagnetism, and limited their discussion mostly to prospecting methods, while many geologists' curriculum practically left it out altogether. The essential contribution provided by the study of ocean floor magnetic anomalies and by paleomagnetism in the development of global tectonic models, made geomagnetism popular in the geological community, which nonetheless continued, and still continues, to view it as a highly specialist discipline.

The authors of this book, like many of their colleagues, are convinced that geomagnetism is now an essential part of any Earth scientist's education. For this reason the book is meant as a first step, presenting fundamental concepts and their more and more numerous applications in many fields of geology, and stimulating readers' interest in further studying the subjects they find most interesting in the many available specialist books. Presenting such a complex, wide-ranging subject as geomagnetism in general terms, requires a drastic choice, both in terms of what to write and of how to write it. A selection of subjects will necessarily be influenced by the authors' education; expressing in a simple and thus approximate form physical concepts that should be articulated with due rigor may lead into error. Whatever judgment the readers may ultimately pass on our work, we will deem we have done something useful if, once they are finished reading it, at least some of them will go to a library to consult far more substantial books and browse the vast geomagnetic literature scientific papers.

The first four chapters of the book discuss the fundamental subjects of geomagnetism within geology: the Earth's magnetic field, the magnetic properties of rocks, measuring and interpreting magnetic anomalies, and paleomagnetism. The next four chapters briefly go over other fields of application: the magnetic fabric of rocks, the Earth's crust magnetization, magnetic chronology and environmental geomagnetism. A short historical chapter ends the book.

First of all we would like to thank those who encouraged us to study geomagnetism: our teachers, who passed their precious experience on to us, and those among our students who asked us the awkward, yet essential, questions that require some sort of answer. As it is impossible to thank each and every one of the persons who helped us along, we will have to restrict ourselves to mentioning the most substantial contributions. Elena

Zanella prepared the figures of Chaps. 2 and 4 to 8, combining her geomagnetic knowledge and her graphic skills; Roberta Tozzi drew those of Chaps. 1 and 3. Seb De Angelis
and Katia Damiani helped to put our concepts in a proper English form. Uwe Zimmer-
mann turned the raw manuscript into a finished book. The various chapters benefited
greatly from the comments and suggestions expressed on a preliminary draft by Don
Tarling, David Barraclough, Niels Abrahamsen, Paola De Michelis, Ted Evans, Ann Hirt,
Frantisek Hrouda and Nicolas Thouveny. Enzo Boschi is thanked for his advice and
support. Last, but perhaps foremost in importance, is the Publisher, who had confidence
in our idea and gave us the opportunity to make it real.

Roberto Lanza
Antonio Meloni

Contents

Acknowledgements

The figures listed below were reproduced from or redrawn based on illustrations in journals and books. The original authors are cited in the figure caption and full reference is given in the "Suggested Readings and Sources of Figures" section of each chapter. Every effort has been made to obtain permission to use copyrighted material. The author and publishers listed below are gratefully acknowledged for giving their kind permission, and apologies are rendered for any errors or omissions.

Publisher	Figure number(s)
Academic Press (Elsevier)	1.24, 1.25, 1.26, 1.27, 1.28, 4.39, 8.6
American Association for Advancement of Science	7.1
American Geophysical Union	2.18, 4.6, 4.7, 4.28, 5.11, 5.23, 6.9, 7.2, 7.3, 7.4, 7.5, 8.4, 8.7, 8.10, Plate 1
Annals of Geophysics, INGV	3.9, Plate 2, 8.2
Birkhauser Verlag	5.21
Blackwell Scientific Publications	4.6, 4.22, 4.31, 7.8
Blackie (Springer Science and Business Media)	2.6, 2.11, 2.13
Cambridge University Press	1.10b, 1.22, 3.7, 3.8, 4.4, 4.27, 4.29
Chapmann & Hall (Springer Science and Business Media)	5.3, 5.6
Elsevier	1.17, 2.9, 2.19, 2.20, 3.5, 4.34, 4.37, 5.7, 5.9, 5.10, 5.12, 5.13, 5.16, 5.20, 6.3, 6.9, 6.10, 7.7, 7.9, 7.10, 7.12, 8.6, 8.13, Plate 4
Geological Society of America	6.1, 7.6, Plate 3
IAGA	1.9
Kluwer Academic (Springer Science and Business Media)	4.19, 8.14
Macmillan Publishers Ltd.	4.40, 6.8
Princeton University Press	6.4, 6.6
Springer Science and Business Media	5.15
Taylor & Francis	4.17, 8.3, 8.11, 8.12
Wiley	4.30
Author	Figure number(s)
R. Butler	4.3, 4.32
Websites	Figure number(s)
http://solid_earth.ou.edu/notes/potential/legendre.gif (Copyright 2004, J. Ahern)	1.12
www.worldofrockhounds.com/magnetite.html	Cover photograph
www-geo.phys.ualberta.ca	9.3

Chapter 1

The Earth's Magnetic Field

Our planet is surrounded by a magnetic field (Fig. 1.1). In our experience this phenomenon is revealed for example by a compass needle that points approximately to the north. According to modern geophysical ideas when, at a given point and at a certain time, a measurement of the Earth's magnetic field is carried out, the measured value is the result of the superimposition of contributions having different origins. These contributions can be, at a first glance, considered separately, each of them corresponding to a different source:

a The main field, generated in the Earth's fluid core by a *geodynamo* mechanism;
b The crustal field, generated by magnetized rocks in the Earth's crust;
c The external field, produced by electric currents flowing in the ionosphere and in the magnetosphere, owing to the interaction of the solar electromagnetic radiation and the solar wind with the Earth's magnetic field;
d The magnetic field resulting from an electromagnetic induction process generated by electric currents induced in the crust and the upper mantle by the external magnetic field time variations.

In order to analyze the various contributions, we will start here with the spatial analysis of the most stable part of the Earth's magnetic field (parts a and b), following in particular the procedure used by Gauss who was the first to introduce the analysis of the Earth's magnetic field potential. After this we will describe the Earth's magnetic

Fig. 1.1. Idealized view of the Earth's magnetic field lines of force with Earth represented as a sphere. N and S are the ideal location of the two magnetic poles

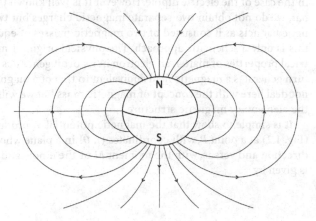

field time variations. In fact the Earth's magnetic field not only shows a peculiar spatial structure, mainly determined by 'a' and 'b' contributions, but is also subject to continuous time variations. These variations, which can have different origins, can be subdivided into two broad classes: long-term and short-term time variations. The former, generally denoted by the name *secular variation*, can be detected when at least 5–10 years, or more, magnetic data from a certain area are examined; this variation is due to the evolution of the deep sources within the Earth, the same sources that also generate the main field. The short-term variations are of external origin to the Earth and are detected over shorter time windows, that can go from fractions of a second generally to no more than a few years (they are essentially included in contribution 'c' above). Finally a magnetic field results from the electromagnetic induction process that is generated by electric currents induced in the crust and the upper mantle, by the external magnetic field time variations. This happens because the Earth is partially an electric conductor and electrical currents can be induced in its conducting parts by external time variations. The secondary magnetic field generated in this way, adds to the other sources.

Only after the results from global analyses of the Earth's magnetic field will be shown, we will give a description of the most important time variations and give an overview of the geodynamo theory. In other chapters the magnetic field of crustal origin and its applications will be discussed.

1.1
Observations and Geomagnetic Measurements

1.1.1
The Magnetic Dipole

The fundamental entity in the study of magnetism is the dipole, that is a system consisting of two magnetic charges, or magnetic masses, of equal intensity and opposite signs. In practice any magnetic bar can be considered a dipole. In some elementary physics books, the term *magnetic mass* is still associated with each end of the dipole. This concept was historically introduced because the magnetic actions exerted by the dipole appear as produced by sources concentrated at its ends, as similarly happens in the case of the electric dipole. However it is well known that if we break a magnetic bar, we do not obtain two separate magnetic charges but two new dipoles. The magnetic bar acts as it consisted of two magnetic masses of equal and opposite signs but this is only a schematic approach. The physical origin of magnetism lies in the electrical properties of matter, an electron in its orbit generates an electric current that in turn generates a magnetic field equivalent to that of a magnetic bar. Therefore we will not deal here with the concept of magnetic mass but we will consider the dipole to be the elementary magnetic structure.

It is simple to show that the magnetic potential V, produced by a magnetic dipole (Fig. 1.2) at a point P, with coordinates (r, θ) in a plane whose polar axis coincides, in direction and versus, with the moment M of the dipole and the origin with its center, is given by

Fig. 1.2. Magnetic dipole field lines of force. The *arrow* indicates the magnetic dipole, r is the vector distance and θ colatitude, as referred to a point P in polar coordinates

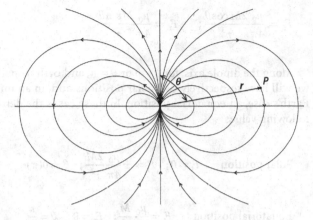

$$V = \frac{\mu_0}{4\pi} \frac{M \cdot r}{r^3} = \frac{\mu_0}{4\pi} \frac{M \cos\theta}{r^2} \tag{1.1}$$

or similarly, using r as a radius vector, the gradient formulation can also be used:

$$V = \frac{\mu_0}{4\pi} M \, grad\left(\frac{1}{r}\right) = \frac{\mu_0}{4\pi} M \nabla\left(\frac{1}{r}\right) \tag{1.2}$$

For symmetry reasons this relationship is valid in all planes passing through the polar axis of the dipole. Therefore in each of these planes we can split the magnetic vector field[1] that we will call F into two components; starting from the following relationship

$$F = -grad V = -\nabla V \tag{1.3}$$

Taking into account the polar coordinate system described above and referring to F_t as the component transverse to the radius vector (positively oriented towards increasing θ, which is called *colatitude*), and to F_r the component directed along r (positively oriented outward), we will obtain

$$F_r = -\frac{\partial V}{\partial r}; \qquad F_t = -\frac{1}{r}\frac{\partial V}{\partial \theta} \tag{1.4}$$

---

[1] $\mu_0 = 4\pi \times 10^{-7}$ Henry m^{-1}, is the magnetic permeability of vacuum. IAGA (International Association of Geomagnetism and Aeronomy) has recommended, as a regular procedure, to use B, magnetic induction, for measurements of Earth's magnetic field instead of H, magnetic field strength.

$$F_r = \frac{\mu_0}{4\pi}\frac{2M\cos\theta}{r^3}; \qquad F_t = \frac{\mu_0}{4\pi}\frac{M\sin\theta}{r^3} \tag{1.5}$$

Along the dipole axis, for $\theta = 0$ or $\theta = \pi$, and orthogonally to this axis, for $\theta = \pi/2$, we will have respectively two polar positions and, in an immediate analogy with the Earth's case, an equatorial position. In these cases the defined components have the following values

Polar position $F_t = 0; \quad F_r = \pm\dfrac{\mu_0}{4\pi}\dfrac{2M}{r^3}; \quad \theta = 0(\pi)$

Equatorial position $F_t = \dfrac{\mu_0}{4\pi}\dfrac{M}{r^3}; \quad F_r = 0; \quad \theta = \dfrac{\pi}{2}$

While for any given value of θ, we have:

$$F = \left(F_t^2 + F_r^2\right)^{\frac{1}{2}} \tag{1.6}$$

$$F = \frac{\mu_0}{4\pi}\frac{M}{r^3}\sqrt{3\cos^2\theta + 1} \tag{1.7}$$

This last relation constitutes the equation of a generic line of force of the dipole magnetic field in polar coordinates.

If the magnetic dipole is immersed in an external magnetic field, as in the case of a magnetic needle or a compass in the Earth's magnetic field, and we let it be free to rotate, both in the horizontal and vertical plane, we can note that it aligns along a particular direction, whatever its original direction was. This is because the needle tends to minimize its interaction energy with the magnetic field in which it is immersed. It is possible to note that to make the interaction energy with an external magnetic field a minimum, a dipole tends to be parallel to a line of force of the external field. If we indicate with F the external magnetic field and with M the dipole (magnetic needle) magnetic moment, the interaction energy E can be expressed as

$$E = -M \cdot F \tag{1.8}$$

while the mechanical couple, Γ, acting on the dipole is

$$\Gamma = M \times F \tag{1.9}$$

The above formulas use the magnetic field F dimensionally as a magnetic induction; we will see that this is considered a standard approach for the Earth's magnetic field. In geomagnetism most of the theoretical studies and data analyses have been devoted to the reconstruction of the configuration of the lines of force of the Earth's magnetic field.

A noticeable analogy can be made between a simple dipole and the source of the Earth's magnetic field. In fact the first analyses carried out by Gauss, already in the first half of 19th century, confirmed the early Gilbert statement that the Earth's magnetic field, in first approximation, appears as generated by a huge magnetic dipole. This dipole is located, inside the Earth, at its center, and has its axis almost parallel to the axis of Earth's rotation. In order to match the orientation of a magnetic needle with its magnetic north pointing to geographic north, the Earth's dipole moment must be oriented in the opposite direction with respect to the Earth's rotation axis (see Fig. 1.2).

1.1.2
Elements of the Earth's Magnetic Field

From now on we will denote by F the Earth's magnetic field vector and, even though currently called a magnetic field, it is intended as a magnetic induction field, which in common physics text books is referred to as B. It can be decomposed on the Earth's surface, along three directions. Considering the point of measurement as the origin of a Cartesian system of reference, the x-axis is in the geographic meridian directed to the north, y-axis in the geographic parallel directed to the east and z-axis parallel to the vertical at the point and positive downwards. The three components of the Earth's magnetic field along such axes are called X, Y and Z (Fig. 1.3). We will then have

$$\sqrt{X^2 + Y^2 + Z^2} = F; \quad \sqrt{X^2 + Y^2} = H \qquad (1.10)$$

where we have also included H as the horizontal component. In order to describe the field, in addition to the intensive components, we can also use angular elements. They

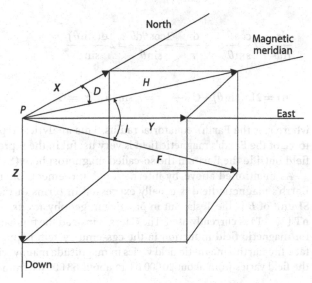

Fig. 1.3. Elements of the Earth's magnetic field. At point P, on the Earth, three axes point respectively to north geographic (x), east geographic (y), and along the vertical downwards (z). The Earth's magnetic field vector F can be projected along the three axes and three magnetic components are obtained X, Y and Z. F also forms an angle I, inclination, with the horizontal plane; H is the horizontal projection of F and angle D, declination, is the angle between H and X

are obtained by introducing two angles, that is I, the inclination of vector F with respect to the horizontal plane, and D declination, the angle between H, the horizontal component of F, and the X component, along the geographic meridian. The relationships among these quantities now defined, are

$$H = F\cos I; \quad Z = F\sin I; \quad Z = H\tan I; \quad X = H\cos D; \quad Y = H\sin D \tag{1.11}$$

Three of these quantities (provided they are independent of each other) are completely sufficient to determine the Earth's magnetic field. Note that H is in geomagnetism the horizontal component of F and must not be confused with the generally agreed use of H in physics where the magnetic field strength is in general intended.

Representing the Earth as a sphere and assuming in first approximation that the field is generated by a dipole placed at its center and pointing towards a given direction, we can visualize a new geometry. The dipole axis through the Earth's center, can be called a *geomagnetic axis* and we obtain that, at a point P on the Earth's surface, what in the magnetic dipole geometry previously were indicated by F_t and F_r, now are equivalent to the horizontal and vertical components of the geomagnetic dipole field:

$$H = F_t; \quad Z = -F_r; \quad \sqrt{H^2 + Z^2} = F^2 \tag{1.12}$$

In such a dipole field the geometry of the lines of force, which will be denoted by the function $r = r(\theta)$, can be derived from

$$\frac{Z}{H} = \frac{dr}{rd\theta} = 2\frac{M\cos\theta}{M\sin\theta} = 2\cot\theta$$

and moreover

$$\frac{dr}{rd\theta} = 2\frac{\cos\theta}{\sin\theta}; \quad \frac{dr}{r} = \frac{2\cos\theta d\theta}{\sin\theta} = 2\frac{d(\sin\theta)}{\sin\theta}$$

$$\ln r = 2\ln(\sin\theta) + C \quad \rightarrow \quad r = r_e\sin^2\theta; \quad r_e = C$$

where r_e is the Earth's equatorial radius. This analytical representation of the lines of force of the Earth's magnetic field is very useful in the representation of the magnetic field outside the Earth in the so-called magnetosphere (Sect. 1.3.4.1).

As mentioned above, by international agreement, the measurement unit for the Earth's magnetic field is usually expressed in terms of the induction vector B. The SI unit of B is the Tesla, but in practice in geophysics its submultiple, the nanoTesla, nT (10^{-9} T) is currently used. The Gauss is instead the fundamental unit of measurement for magnetic field induction in the cgs-emu system (Appendix). On the Earth's surface the Earth's magnetic field varies in magnitude mainly with latitude; to grab an idea, the field varies from about 20 000 nT to about 68 000 nT from the equator to the poles. In

Figs. 1.4, 1.5 and 1.6, the horizontal, vertical and total magnetic field isodynamic charts showing the spatial variations of the given element on the Earth's surface for the year 2005, are reported; in Fig. 1.7 the isogonic map for declination at year 2005 is reported.

1.1.3
Early Measurements of the Earth's Magnetic Field

The object of geomagnetic measurements is the quantitative determination of the Earth's magnetic field elements; this is done using magnetic instruments, called *magnetometers*. Over the years many kind of magnetometers have been designed by scholars and specialists in order to improve the quality of the measurement or to reach a better portability, efficiency, or ease of use. We will not go all the way through the long history of magnetic instruments here, we will however start with a brief introduction describing classical mechanical magnetometers and then we will directly proceed with the more modern and widely used instruments, based on electromagnetic or nuclear phenomena, which make a large use of modern electronics.

Gauss was the first to construct a complete set for the absolute determination of the geomagnetic field elements in the early years of the 19th century. Being the geomagnetic field a vector it is in fact self evident that its complete determination needs the quantification of all elements of this vectorial quantity. The magnetic compass was already used in the middle ages employing magnetic needles to point the magnetic north. The almost faithful north indication made the compass a very useful instrument for north bearing, especially for ships. Around the 15th century it became clear that the compass was not pointing precisely to the geographic north but that an angle, later on called declination, was separating magnetic north indication from geographic north indication. So by an independent measurement of the geographic north, a magnetic needle mounted on a horizontal circle allowed the determination of the declination angle in the horizontal plane. The inclinometer, probably introduced during the 16th century, gives the magnetic field F inclination with respect to the horizontal plane. Inclinometers also used magnetic needles but the needle was pivoted around a horizontal axis;

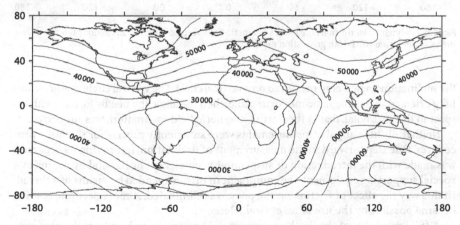

Fig. 1.4. Isodynamic world chart for Earth's magnetic total field F. Contour lines in nT, for the year 2005 from IGRF 10th generation model

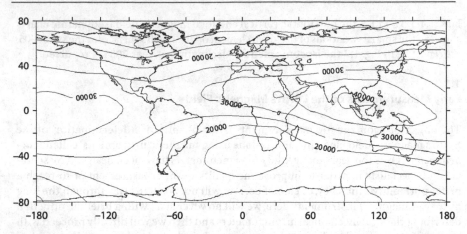

Fig. 1.5. Isodynamic world chart for Earth's magnetic field horizontal intensity *H*. Contour lines in nT, for the year 2005 from IGRF 10th generation model

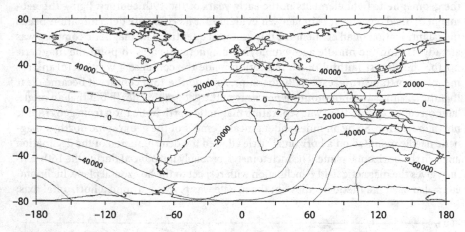

Fig. 1.6. Isodynamic world chart for Earth's magnetic field vertical intensity *Z*. Contour lines in nT, for the year 2005 from IGRF 10th generation model

the inclination angle being measured on a vertical circle. The vertical circle was first carefully oriented in the magnetic meridian plane, then the angle the needle formed with respect to the horizontal, that is the Earth's magnetic field inclination, was measured.

Neither of these angular measurements were sufficiently precise for scientific procedures. One step forward in the measurement of declination, improving its accuracy, was made with the introduction of suspended needles, kept horizontal by means of a special supporting equipment, the equipment in turn suspended by means of a thread. In this way the effect of friction on the pivot was eliminated. A more accurate reading became possible by the use of an optical telescope.

A full knowledge of the Earth's magnetic field vector *F* needs at least the measurement of one of its intensive components. The well known explorer Von Humboldt used

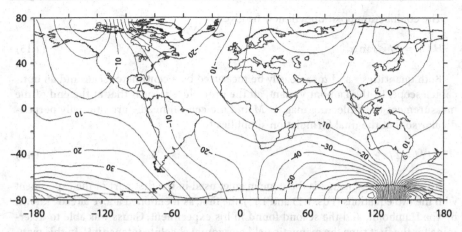

Fig. 1.7. Isogonic world chart for Earth's magnetic field declination D. Contour lines in degrees (°), for the year 2005 from IGRF 10th generation model

the observation of the time of oscillation of a compass needle in the horizontal plane to determine relative measurements of horizontal intensity using the relation

$$T = 2\pi \, (I / MH)^{1/2} \tag{1.13}$$

which connects, for small amplitude oscillations, the period of oscillation T of a magnet with its moment of inertia I and magnetic moment M in a horizontal magnetic field H. This very simple method reduced relative H measurements to the measurement of the oscillation period of a magnet. The procedure was generally adopted by several observers in scientific journeys allowing to obtain a first order approximation knowledge on magnetic field magnitude variation around the globe. Unfortunately in order to establish the absolute magnitude of the magnetic field H, the determination of the needle magnetic moment M and moment of inertia I was necessary.

In 1832 Gauss was the first to realize that it was possible to devise a procedure for the correct absolute determination of the Earth's magnetic field horizontal intensity. This method, modified later by Lamont, consists in the comparison of two mechanical couples acting on a horizontal suspended magnetic needle. One couple is the Earth's magnetic field couple, while the second is artificially acted by a magnet located at a fixed distance from the oscillating needle. In a first phase of the measurements the magnetic needle is accurately oriented along the Earth's magnetic field; in a second phase a deflecting magnet is put in operation at a distance r, laterally at a right angle to the central needle. Calling M the deflecting magnet magnetic moment, the central needle will experience not only the Earth's magnetic field horizontal intensity, H, but also a second field, whose intensity we can call H_1, generated by the deflecting magnet M:

$$H_1 = 2M / r^3 \tag{1.14}$$

As a result the central needle (Fig. 1.8) will be under the influence of the two couples which will move it to a new position, forming an angle α with the initial direction.

The equilibrium position will now be given by

$$H / M = 2 / r^3 \tan \alpha \qquad (1.15)$$

Both quantities r and α can easily be measured by a centimeter scale and an optical telescope. In the Lamont variant, all the procedure is such that at the end of the measurement the deflecting magnet M and the central needle are mutually perpendicular so that the final formulation simplifies to

$$H / M = 2 / r^3 \qquad (1.16)$$

If the deflecting magnet is the same magnet used in the first part of the experiment with the two equations (Eqs. 1.13 and 1.15), the first, as mentioned above already known by Von Humboldt, and the second found in his experiment, Gauss was able to determine for the first time the magnetic field horizontal absolute intensity H. In this manner the Earth's magnetic field became the first non-mechanical quantity expressed in

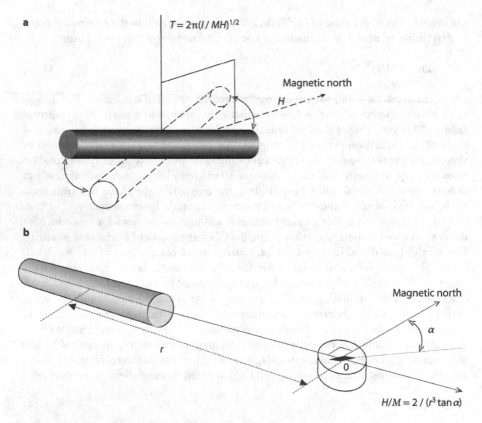

Fig. 1.8. Gauss Lamont magnetometer; **a** a magnetic bar oscillates with period T in the Earth's magnetic field; **b** the magnetic bar is now used to deflect a magnetic needle that rotates freely to an equilibrium position in the magnetic bar and the Earth's magnetic fields

terms of the three fundamental mechanical quantities: mass, length and time. This result was reported in the Gauss's memoir *Intensitas vis Magneticae Terrestris ad Mensuram Absolutam Revocata* in 1833, the last great scientific memoir written in Latin. The complete instrument used in his procedure was for the first time called a magnetometer.

1.1.4
Modern Magnetic Measurements

Since the Earth's magnetic field is a vector quantity, the field magnitude is absolute if expressed in terms of the fundamental quantities (for example mass, length, time and electrical current intensity), while the vector spatial orientation can be expressed for example in terms of D and I, angular dimensionless quantities. From the total field F magnitude and the angular quantities, the geomagnetic field components H, Z and also X, Y can be computed. Sometimes magnetic instruments give as outputs directly the geomagnetic components; it is self evident that once three independent elements are determined, the magnetic field measurement is considered complete.

Nowadays magnetic instruments that utilize magnets for their operation are only very seldom used in magnetic observatories. Moreover the measurement of declination and inclination angles is a procedure employed mainly for absolute magnetic measurements in magnetic observatories or at repeat magnetic stations. An instrument is called absolute when it gives the value of the measured quantity in terms of one or more of the absolute basic fundamental quantities of physics. For this reason in geomagnetism the term *absolute measurement* is still often used to indicate a procedure for the complete absolute determination of the magnetic field elements. An instrument is called relative when it measures the value of one element of the Earth's field as a deviation from a certain initial value not necessarily known. Many of these instruments require a reference initial value that must be determined independently, for example by means of an absolute instrument. The use of relative instruments can of course be very convenient especially in some field operations, for example when only the spatial variation of the magnetic field in an investigated area is required. A second case is when, at a given place, a time variation of the Earth's magnetic field needs to be recorded.

Instruments are delivered with information and data sheets that provide the values of the parameters necessary to evaluate their measurement capability. The most frequently used parameters are reported in what follows.

- *Accuracy:* indicates how an instrument is accurate, that is the maximum difference between measured values and true values.
- *Precision:* is related to the scatter of the measured values and refers to the ability of the instrument of repeating the same value when measuring the same quantity.
- *Resolution:* represents the smallest change of the measured quantity that is detectable by the instrument.
- *Range:* refers to the upper and lower (extreme) limits that can be measured with the instrument. The dynamic range is the ratio between the maximum measurable quantity and the resolution, normally expressed in dB, i.e. $20\log(A_{max}/A_{min})$.

- *Sensitivity:* indicates how many scale units of the instrument correspond to one unit of the measured physical unit.
- *Scale value:* is the reciprocal of sensitivity.

Magnetic instruments are nowadays not only devoted to magnetic measurement, they are also frequently equipped with electronic cards, able to memorize measured data and to interface to PCs for real-time or off-line data communication.

1.1.4.1
Absolute Instruments

Proton Precession Magnetometers and Overhauser Magnetometers
These instruments are based upon the nuclear paramagnetism, i.e. the circumstance that atomic nuclei posses a magnetic spin that naturally tends to orient itself along an external magnetic field. In these magnetometers the sensor is made up of a small bottle full of a hydrogenated liquid (such as propane, decane or other that can operate as liquid in a reasonable temperature range) around which a two coil system is wounded. A direct electrical current is applied to the first winding (polarization coil) by means of an external power supply and consequently generates a magnetic field inside the bottle. Protons in the bottle are then forced to align their spin along this magnetic field starting to precess at a frequency rate depending on the magnetic field magnitude. If the external current is interrupted, the artificial magnetic field is removed and then protons in the bottle will start precessing around the Earth's magnetic field direction at a frequency f given by

$$f = (\gamma / 2\pi)F \tag{1.17}$$

where γ is the so-called magneto-mechanical proton ratio (gyromagnetic ratio) a fundamental quantity, very precisely known in atomic physics (2.6751525×10^8 rad s^{-1} T^{-1}) and F is the external Earth's magnetic field. The proton precession generates at the ends of the second winding (pick-up coil) a time varying electromotive force (e.m.f.) with the same frequency, which can easily be measured to obtain the absolute total field F magnitude. In the average Earth's magnetic field (for example 45 000 nT) the frequency is very close to 2 kHz (1 916 Hz) (Fig. 1.9).

The loss of coherence inside the bottle allows only a small time window (about 2–3 s) for the detection of the e.m.f. frequency. This time is however now more than sufficient for modern electronic frequency meters to give the precession frequency. In fact due to progress in electronic technology, the measurement of frequency is in contemporary physics one of the most accurate techniques. Since it is only dependent on the measurement of a frequency, the measurement of the Earth's magnetic field by means of a proton precession magnetometer is both very precise and absolute: resolution reaches now easily 0.1 to 0.01 nT.

One disadvantage of proton precession magnetometers is the limitation due to the fact that the polarization current needs to be switched off in order to make a measurement. The operation is therefore discontinuous with a time interval of a few seconds between measurements. A continuous proton precession signal can however be obtained by taking advantage for example of the so-called Overhauser effect. The ad-

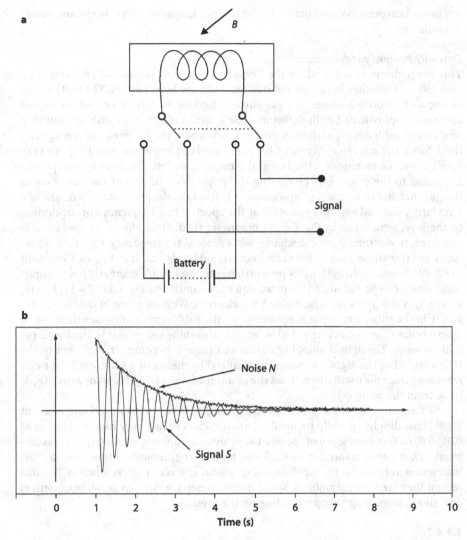

Fig. 1.9. Proton precession magnetometer; **a** electric circuitry schematics for measurement of field B. The measurement is performed in two steps: (1) generation of free proton precession by power injection; (2) signal detection after switching; **b** typical detected signal amplitude decrease. Signal to noise ratio is optimum for only a few seconds after polarization is turned off (from Jankowsky et al. 1996)

dition of free electrons into the liquid in the bottle and the application of a suitable radio frequency, can in fact increase the magnetization of the liquid sample. Without going into details, we will just remember here that as an alternative to applying a strong polarizing field, in Overhauser magnetometers the magnetization is increased by applying a suitable radio frequency electromagnetic field to put the free electrons into resonance. This electron resonant frequency that exceeds by 658 times the proton resonant frequency, has the role of increasing the proton level saturation making the proton precession process signal output in Overhauser magnetometers continuous rather than discrete.

Optically Pumped Magnetometer
This magnetometer is based on the Zeeman effect and the so-called stimulated emission of radiation in certain substances, as in the Maser effect. The instrument consists of a bottle containing a gas such as helium, rubidium or cesium vapors and some sophisticated light detectors. The Zeeman effect deals with the splitting of electron sublevels separation in energy levels under the influence of a magnetic field. Since the energy differences between levels of hyperfine splitting are very small, a specific technique, called optical pumping, is used. The term optical pumping refers to the process of increasing the population of one of the sublevels in the gas that, in the measuring procedure is initially underpopulated, by means of a circularly polarized external radiation at the spectral line frequency corresponding to the level separation in the Earth's magnetic field. Once the overpopulation is obtained, an electromagnetic discharge takes place at the frequency $f = \Delta E / h$ where ΔE is the transition energy between Zeeman sublevels and h is Planck's constant (6.62×10^{-34} joule–second). ΔE is proportional to the Earth's magnetic field magnitude, which can be calculated from an equation similar to Eq. 1.17: $f = (\gamma_e / 2\pi)F$, where γ_e is the gyromagnetic ratio of the electron. Without going in details, for example in the rubidium vapor magnetometer, the rubidium light passes through an interference filter, a circular polarizer and an absorbing cell, which is filled with rubidium vapor. The light is pumping atoms and the cell becomes transparent to the resonant light. The light intensity is measured by means of a photocell; the radio frequency from the oscillator follows the resonance frequency due to negative feedback from the photocell.

The frequency f is of the order of 200 kHz and γ_e is known to a precision of about $1/10^7$. Accordingly, optically pumped magnetometers have a very high resolution of 0.01–0.001 nT and are some of the most sensitive instruments for magnetic measurements. Their performance can be exploited almost continuously in time, making this instrument very useful for rapid data acquisition at a very high resolution. For this reason they are very common in space magnetometry as well as in aeromagnetism and also in some magnetic prospecting on the ground.

1.1.4.2
Relative Instruments

Fluxgate Magnetometers
Fluxgate magnetometers are electromagnetic instruments that can give direct magnetic measurements along a built-in direction. By orienting this direction along the Earth's magnetic field elements such as F, Z or H, these elements may be measured. The orienting device may vary according to requirements. In some fluxgate magnetometers this built-in direction is along a straight cylinder, while in others the direction is taken along the plane of a ring shaped sensor.

In one class of fluxgate magnetometers the sensor unit is constituted by a cylindrical core with very high magnetic permeability (for example made of permalloy, mu-metal or ferrite) placed inside two windings. In the first winding a 1 000 Hz excitation current flows and generates an alternate magnetic field, large enough to saturate the core. In the absence of an external steady magnetic field acting on the sensor, such as a component of the Earth's field, the alternating field collected by the second winding (pick-up coil) contains only the odd harmonics of the excitation current. If a steady magnetic field acts along the core axis, then this field sums to the alternating one in such a way that one semiwave of the e.m.f. in the pick-up coil is now larger in amplitude than the one generated in the opposite direction. In this case in fact the core is brought to saturation faster in the direction parallel to the Earth's field than in the opposite direction. In the pick-up coil a double frequency current will now appear; the amplitude of this current is linearly proportional to the magnitude of the external field acting along the core direction (Fig. 1.10).

In other fluxgate magnetometer models, the central core is substituted by two parallel ferromagnetic cores arranged in such a way that the alternating current acting on the cores produces at the excitation winding terminals two equal and out of phase e.m.f. exactly balanced, which thus sum up to zero. When an external magnetic field acts on the cores, this symmetry is broken and the varying e.m.f. induced in the pick-up coil is linearly proportional to the magnitude of the external field.

In actual fluxgate magnetometers the sensor excitation is produced by means of an electronic oscillator, the signal from the pick-up coil is fed into a tuned amplifier and the output is fed to a phase sensitive detector referenced to the second harmonic of the excitation frequency. The fluxgate magnetometer is a zero field instrument. This means that in order to measure the full intensity of the geomagnetic field along one of its components it also needs an auxiliary compensation system. One serious problem in fluxgate magnetometers, is the temperature variation; in fact the bias coils need a stabilization. To obtain this stabilization the coils have to be wound around quartz tubes or other thermally stable material frames. The fluxgates have a reasonable 0.1 nT resolution and are non absolute instruments frequently used for recording magnetic time variations.

Fig. 1.10a. Fluxgate magnetometer; winding schematics in the case of a two core fluxgate instrument

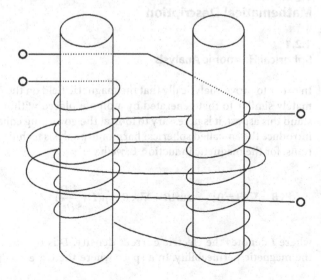

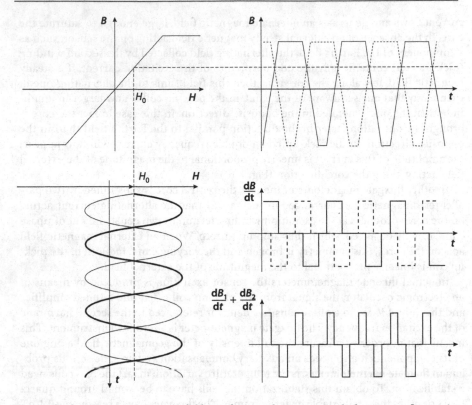

Fig. 1.10b. Fluxgate magnetometer; B field waveforms at the output signal of a two core fluxgate instrument (from Lowrie 1997)

1.2
Mathematical Description

1.2.1
Spherical Harmonic Analysis

In order to prove analytically that the magnetic field on the Earth's surface is approximately similar to that generated by a dipole placed within the Earth, and to understand this aspect, it is necessary to look at the governing equations for magnetism and introduce the so-called spherical harmonic analysis technique. From Maxwell's equations, for the magnetic induction B, we have

$$\text{div}B = \nabla \cdot B = 0; \quad \text{curl}B = \nabla \times B = \mu(I + \frac{\partial D}{\partial t}) \tag{1.18}$$

where I denotes the electric current density, D is the dielectric induction and μ is the magnetic permeability. In a space where there are no discontinuity surfaces and no electric currents, it can be assumed that B can be derived from a magnetic potential V

$$\text{curl} B = 0; \quad B = -\text{grad} V$$

$$\text{div}(-\text{grad} V) = \nabla^2 V = 0$$

$$\Delta V = 0$$

where Δ is the so-called Laplacian operator equivalent to ∇^2. The last equation for V, known as Laplace's equation, in orthogonal Cartesian coordinates, becomes

$$\Delta V = \frac{\partial^2 V}{\partial x^2} + \frac{\partial^2 V}{\partial y^2} + \frac{\partial^2 V}{\partial z^2} = 0 \tag{1.19}$$

and can be written in a spherical coordinate system, with the origin at the Earth's center, as

$$\Delta V = \frac{1}{r^2}\frac{\partial}{\partial r}\left(r^2 \frac{\partial V}{\partial r}\right) + \frac{1}{r^2 \sin\theta}\frac{\partial}{\partial \theta}\left(\sin\theta \frac{\partial V}{\partial \theta}\right) + \frac{1}{r^2 \sin^2\theta}\frac{\partial^2 V}{\partial \lambda^2} = 0 \tag{1.20}$$

Each function $V = V(r, \theta, \lambda)$ that satisfies this equation is called a harmonic function; r and θ are defined as in Sect. 1.1.1 where the dipole geometry was described in a given plane and let now λ be the geographic longitude, we can assume the Earth to ·be a sphere of radius a (Fig. 1.11).

Fig. 1.11. Earth-centered coordinates x, y and z with origin at the Earth's center. Spherical coordinates: for point P on the Earth's surface, r is the distance from Earth's center, θ colatitude and λ longitude

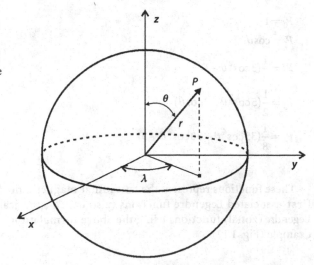

The general solution for a potential in Laplace's equation can be obtained (as similarly occurs in the Earth's gravity case), by means of a technique called spherical harmonic analysis (SHA). The determination of three orthogonal functions, expressed in terms of only one variable each, is needed. In the search for these functions we will take into account the characteristics of the field as considered in spherical coordinates.

Starting from the variable r those functions that take into account the two possible origins of the field, internal or external to the Earth respectively, are considered. As regards λ its definition demands a periodic behavior from 0 to 2π, suggesting the use of periodic functions, as in case of a Fourier series in λ. For what concerns θ, in geomagnetism, the Schmidt quasi-normalized functions are used; similarly to the case of gravity potential solutions, these are equivalent to Legendre functions $P_{n,m}(\theta)$, but with a different normalizing factor. Schmidt functions are in fact normalized to be of the same order of magnitude as the zonal Legendre functions of the same degree. Let us first refer to usual Legendre functions

$$P_{n,m}(\theta) = \sin^m \theta \frac{d^m P_n(\theta)}{d(\cos\theta)^m} \tag{1.21}$$

where n denotes the degree and m is the order, and, for $m = 0$, these reduce to the standard zonal functions

$$P_n(\theta) = \frac{1}{n!2^n} \frac{d^n}{d(\cos\theta)^n} (\cos^2\theta - 1)^n \tag{1.22}$$

Only as an example we recall the first few Legendre zonal functions $P_n(\theta)$.

$$P_0 = 1$$
$$P_1 = \cos\theta$$
$$P_2 = \frac{1}{2}(3\cos^2\theta - 1)$$
$$P_3 = \frac{1}{2}(5\cos^3\theta - 3\cos\theta)$$
$$P_4 = \frac{1}{8}(35\cos^4\theta - 30\cos^2\theta + 3)$$

These functions represent the latitudinal magnetic field variations. Some of the first associated Legendre functions (also called spherical functions), drawn from Legendre (zonal) functions using the above formulation, are also shown here as an example (Fig. 1.12).

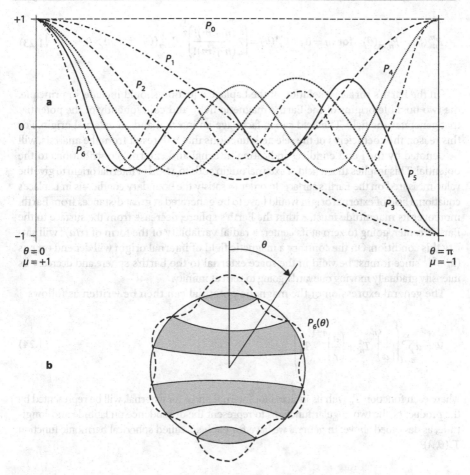

Fig. 1.12. A few low-degree (P_0–P_6) Legendre zonal harmonics on the Earth surface shown for $0 < \theta < \pi$. In the *lower part* function P_6 is shown along the circumference of a circle with gray and white zones to indicate negative and positive values for P_6 on the spherical surface (from Ahern, Copyright 2004)

$$P_1^1 = \sin\theta$$
$$P_2^1 = 3\cos\theta\sin\theta$$
$$P_2^2 = 3\sin^2\theta$$
$$P_3^3 = 15\sin^3\theta$$
$$P_4^4 = 105\sin^4\theta$$

The Schmidt functions used in geomagnetism, indicated by $P_n^m(\theta)$, are partially normalized Legendre functions, differing from Legendre functions only by a normalizing factor. They are defined as follows:

$$P_n^m(\theta) = P_{n,m}(\theta) \quad \text{for } m = 0; \quad P_n^m(\theta) = \left(2\frac{(n-m)!}{(n+m)!}\right)^{\frac{1}{2}} P_{n,m}(\theta) \quad \text{for } m > 0 \tag{1.23}$$

On the Earth's surface a solution of the Laplace equation should include, in principle, the two possible origins of the Earth's magnetic field, and consequently of the potential that generates the field. The field can in fact have an internal and an external origin. For this reason the coefficients of the selected functions that will result from the analysis, will be denoted by the i and e indices indicating internal and external contributions to the potential. This implies that a field of external origin adds to a field of internal origin to give the value measured on the Earth's surface. In order to satisfy the boundary conditions in Laplace's equation, a field of external origin would have to be generated at great distances from Earth, moreover its magnitude must, within the Earth's sphere, decrease from the surface to the Earth's center, going to zero at its center: a radial variability of the form of $(r/a)^n$ will satisfy this condition. On the contrary a magnetic field of internal origin will depend on r as $(a/r)^{n+1}$ since it must be valid in the space external to the Earth's sphere and decrease its intensity gradually moving outwards, going to zero at infinity.

The general expression of the magnetic potential can then be written as follows

$$V = a \sum_{n=1}^{\infty} \left(\left(\frac{r}{a}\right)^n T_n^e + \left(\frac{a}{r}\right)^{n+1} T_n^i \right) \tag{1.24}$$

where each function T_n with its indices i for internal and e for external, will be represented by the product of the two angular functions to represent the dependence on latitude and longitude, as described above. In general we have for each so-called spherical harmonic function $T_n(\theta, \lambda)$

$$T_n = \sum_{m=0}^{n} (g_n^m \cos m\lambda + h_n^m \sin m\lambda) P_n^m(\theta) \tag{1.25}$$

where g_n^m and h_n^m are the expansion coefficients for the magnetic potential, traditionally called in geomagnetism the Gauss coefficients.

Since the quantities on which an actual data analysis can be undertaken are the magnetic field components, the mathematical form of these quantities, that is the components of the vector field, must be considered. They are the space derivatives of the potential function V defined above:

$$X = \frac{\partial V}{r \partial \theta} = \frac{a}{r} \sum_{n=1}^{\infty} \sum_{m=0}^{n} \frac{\mathrm{d}P_n^m}{\mathrm{d}\theta} \left[\begin{array}{l} \left\{ g_n^{me} \left(\frac{r}{a}\right)^n + g_n^{mi} \left(\frac{a}{r}\right)^{n+1} \right\} \cos m\lambda \\[2mm] + \left\{ h_n^{me} \left(\frac{r}{a}\right)^n + h_n^{mi} \left(\frac{a}{r}\right)^{n+1} \right\} \sin m\lambda \end{array} \right]$$

$$Y = \frac{-\partial V}{r\sin\theta\partial\lambda} = \frac{a}{r\sin\theta}\sum_{n=1}^{\infty}\sum_{m=0}^{n}P_n^m\left[\begin{array}{l}m\left\{g_n^{me}\left(\dfrac{r}{a}\right)^n + g_n^{mi}\left(\dfrac{a}{r}\right)^{n+1}\right\}\sin m\lambda \\[2mm] -m\left\{h_n^{me}\left(\dfrac{r}{a}\right)^n + h_n^{mi}\left(\dfrac{a}{r}\right)^{n+1}\right\}\cos m\lambda\end{array}\right]$$

$$Z = \frac{\partial V}{\partial r} = \sum_{n=1}^{\infty}\sum_{m=0}^{n}P_n^m\left[\begin{array}{l}\left\{ng_n^{me}\left(\dfrac{r}{a}\right)^{n-1} - (n+1)g_n^{mi}\left(\dfrac{a}{r}\right)^{n+2}\right\}\cos m\lambda \\[2mm] +\left\{nh_n^{me}\left(\dfrac{r}{a}\right)^{n-1} - (n+1)h_n^{mi}\left(\dfrac{a}{r}\right)^{n+2}\right\}\sin m\lambda\end{array}\right]$$

On the Earth's surface, for $r = a$, these equations can be simplified and become

$$X = \sum_{n=1}^{\infty}\sum_{m=0}^{n}\frac{dP_n^m}{d\theta}\left(g_n^m\cos m\lambda + h_n^m\sin m\lambda\right) \tag{1.26}$$

$$Y = \frac{1}{\sin\theta}\sum_{n=1}^{\infty}\sum_{m=0}^{n}P_n^m\left(mg_n^m\sin m\lambda - mh_n^m\cos m\lambda\right) \tag{1.27}$$

$$Z = \sum_{n=1}^{\infty}\sum_{m=0}^{n}P_n^m\left[\left(ng_n^{me} - (n+1)g_n^{mi}\right)\cos m\lambda + \left(nh_n^{me} - (n+1)h_n^{mi}\right)\sin m\lambda\right] \tag{1.28}$$

We immediately note that in the equations for X and Y, the terms of the Gauss coefficients are now expressed as $g_n^m = g_n^{me} + g_n^{mi}$ and $h_n^m = h_n^{me} + h_n^{mi}$. They appear only as the sum of the internal and external contributions making in this way the separation of the two contributions impossible. On the contrary in the equation for Z these terms are still separated. This consideration will let us estimate such contributions separately. It is important to note, however, that from the distribution of measurements of the X and Y components on the Earth's surface, we can obtain two independent evaluations of g_n^m and h_n^m.

1.2.2
Methods for g_n^m and h_n^m Computation

In the last twenty years or so new methods of analysis have brought to sophisticated procedures for the computation of Gauss coefficients from magnetic measurements. For many years however a very simple and intuitive procedure was used.

Firstly isodynamic maps of the X and Y components were drawn for the whole Earth, thanks to a series of values of the field measured at an irregular distribution of points: observatories, field stations and also ship logs. Then from the obtained maps, values of the magnetic field components on a regular net of points, for instance at the crossing of the meridians and the parallels every $10°$ of latitude and longitude, could be obtained by interpolation. Fixing our attention on a particular fixed value of colatitude $\theta = \theta_0$ (parallel) and considering all values of the X component on this parallel, a simple procedure enabled the determination of Gauss coefficients. In fact let us start, for example, from the expansion in spherical harmonics for X (Eq. 1.26). On the array of measured points, on that parallel, let us apply a Fourier expansion (denoted by f) to the real data obtained for the component under consideration

$$f(\theta) = \sum_{m=0}^{n} (a_m \cos m\lambda + b_m \sin m\lambda) \tag{1.29}$$

whose coefficients for a total length of the parallel L, are generally denoted by

$$a_m = \frac{1}{L} \int_{-L}^{L} f(\theta) \cos \frac{m\pi\theta}{L} \, d\theta$$

$$b_m = \frac{1}{L} \int_{-L}^{L} f(\theta) \sin \frac{m\pi\theta}{L} \, d\theta$$

In order to make these two expansions identical, i.e. $f(\theta)$ were equal to $X(\theta_0)$, at $\theta = \theta_0$, the coefficients of the two respective series expansions must be equal, in detail:

$$X = \frac{dP_1^0}{d\theta} g_1^0 + \frac{dP_1^1}{d\theta} g_1^1 \cos\lambda + \frac{dP_1^1}{d\theta} h_1^1 \sin\lambda + \frac{dP_2^0}{d\theta} g_2^0 + \frac{dP_2^1}{d\theta} g_2^1 \cos\lambda + \frac{dP_2^1}{d\theta} h_2^1 \sin\lambda$$
$$+ \frac{dP_3^0}{d\theta} g_3^0 + \dots$$

$$f(\theta) = a_0 + a_1 \cos\lambda + b_1 \sin\lambda + \dots$$

Once the Fourier expansion coefficients are known, this allows us to determine the Gauss coefficients for that component (on the given parallel). The above procedure, referred to the X component, can be similarly applied to the Y component. According to this analysis we have that, if the field is derivable from a potential, the coefficients g_n^m and h_n^m obtained by the two component analysis must be equal within measurement errors.

1.2.3
Results of Spherical Harmonic Analysis

From all data analyses made up to now, it has resulted that the differences between the two sets of coefficients g_n^m and h_n^m obtained independently by the analysis of X and Y components separately, are very small and can be only attributed to measurement errors. The spherical harmonic analysis of the Earth's magnetic field confirms that the assumptions underlying the derivation of Laplace's equation are correct and by consequence the assumption that under certain limitations the Earth's magnetic field is conservative, is valid.

In order to determine separately the contributions of the fields of external and internal origin to the Earth, it is necessary to analyze also the distribution of the vertical component Z. Remembering the formula (Eq. 1.28) obtained for Z from Laplace's equation solution

$$Z = \sum_{n=1}^{\infty} \sum_{m=0}^{n} P_n^m(\theta) \left[\left(n g_n^{me} - (n+1) g_n^{mi} \right) \cos m\lambda + \left(n h_n^{me} - (n+1) h_n^{mi} \right) \sin m\lambda \right]$$

We can define now

$$C_n^m A_n^m = g_n^{me}; \quad (1 - C_n^m) A_n^m = g_n^{mi}; \quad S_n^m B_n^m = h_n^{me}; \quad (1 - S_n^m) B_n^m = h_n^{mi}$$

from which $g_n^{me} + g_n^{mi} = g_n^m = A_n^m$; $h_n^{me} + h_n^{mi} = h_n^m = B_n^m$.
The expression for Z becomes

$$Z = \sum_{n=1}^{\infty} \sum_{m=0}^{n} P_n^m(\theta) \left[\begin{array}{l} \left(nC_n^m - (n+1)(1-C_n^m) \right) g_n^m \cos m\lambda \\ + \left(nS_n^m - (n+1)(1-S_n^m) \right) h_n^m \sin m\lambda \end{array} \right] \tag{1.30}$$

As done for X and Y components, by means of a series of measurements on a regular network, we can approximate the measurements at a given colatitude θ by a periodic trigonometric function. A similar procedure can now be obtained by using the Fourier expansion

$$Z = \sum_{n=1}^{\infty} \sum_{m=0}^{n} P_n^m(\theta)(\alpha_n^m \cos m\lambda + \beta_n^m \sin m\lambda) \tag{1.31}$$

but in this case we must equate the new coefficients α_n^m and β_n^m, deduced in the same way it was done for X and Y components:

$$\alpha_n^m = \left[nC_n^m - (n+1)(1-C_n^m)\right]g_n^m$$

$$\beta_n^m = \left[nS_n^m - (n+1)(1-S_n^m)\right]h_n^m$$

From these equations it is possible to obtain (by the same least mean squares method) α_n^m and β_n^m. Since g_n^m and h_n^m are already known from the horizontal components, we can get the fractions of the harmonic terms respectively of external and internal origin, i.e. C_n^m and S_n^m.

Starting from the analyses carried out since the 1830s from Gauss until today, it results that the terms of external origin show an amplitude far lower, almost negligible, than those of internal origin. In particular as regards the g_1^0 coefficient, the part of external origin is about 0.2% of the internal one, while for g_1^1 and g_1^1 they are about 2%. As it was noted in the case of the comparison between g_n^m and h_n^m, obtained from X and Y (separately computed), we can assume, at this time, that the contribution of external origin is not exactly equal to zero essentially for the following reasons: experimental errors, inability to draw exactly the isodynamic maps and the difficulty to compute the real "mean" magnetic field values within a certain time interval.

In conclusion we can say that the potential, and therefore the Earth's magnetic field, is of internal origin. The potential function of the geomagnetic field can be completely formulated taking only into account the terms of internal origin, denoting from now on Gauss coefficients without the index i

$$V = a \sum_{n=1}^{\infty} \sum_{m=0}^{n} \left(\frac{a}{r}\right)^{n+1} P_n^m(\theta)(g_n^m \cos m\lambda + h_n^m \sin m\lambda) \tag{1.32}$$

A full set of Gauss coefficients for the years 2000 and 2005, up to degree and order $n = m = 6$, is given in Table 1.1.

1.2.4
A Predominantly Dipolar Field

Taking only into account the terms of the potential expansion up to $n = 1$, the expression for V becomes at any point on the Earth's surface $P = P(\theta, \lambda)$

$$V_1 = \frac{a^3}{r^2}(g_1^0 P_1^0 + g_1^1 P_1^1 \cos\lambda + h_1^1 P_1^1 \sin\lambda) \tag{1.33}$$

since $P_1^0 = \cos\theta$ and $P_1^0 = \sin\theta$

$$V_1 = \frac{a^3}{r^2}(g_1^0 \cos\theta + g_1^1 \sin\theta \cos\lambda + h_1^1 \sin\theta \sin\lambda) \tag{1.34}$$

Table 1.1. Gauss coefficients g and h in nT for $n = 1$ to 6, for the years 2000 and 2005 with secular variation coefficients for 2005–2010 in the 10th generation IGRF

g/h	n	m	2000.0	2005.0	SV
g	1	0	−29619.4	−29556.8	8.8
g	1	1	−1728.2	−1671.8	10.8
h	1	1	5186.1	5080.0	−21.3
g	2	0	−2267.7	−2340.5	−15.0
g	2	1	3068.4	3047.0	−6.9
h	2	1	−2481.6	−2594.9	−23.3
g	2	2	1670.9	1656.9	−1.0
h	2	2	−458.0	−516.7	−14.0
g	3	0	1339.6	1335.7	−0.3
g	3	1	−2288.0	−2305.3	−3.1
h	3	1	−227.6	−200.4	5.4
g	3	2	1252.1	1246.8	−0.9
h	3	2	293.4	269.3	−6.5
g	3	3	714.5	674.4	−6.8
h	3	3	−491.1	−524.5	−2.0
g	4	0	932.3	919.8	−2.5
g	4	1	786.8	798.2	2.8
h	4	1	272.6	281.4	2.0
g	4	2	250.0	211.5	−7.1
h	4	2	−231.9	−225.8	1.8
g	4	3	−403.0	−379.5	5.9
h	4	3	119.8	145.7	5.6
g	4	4	111.3	100.2	−3.2
h	4	4	−303.8	−304.7	0.0
g	5	0	−218.8	−227.6	−2.6
g	5	1	351.4	354.4	0.4
h	5	1	43.8	42.7	0.1
g	5	2	222.3	208.8	−3.0
h	5	2	171.9	179.8	1.8
g	5	3	−130.4	−136.6	−1.2
h	5	3	−133.1	−123.0	2.0
g	5	4	−168.6	−168.3	0.2
h	5	4	−39.3	−19.5	4.5
g	5	5	−12.9	−14.1	−0.6
h	5	5	106.3	103.6	−1.0
g	6	0	72.3	72.9	−0.8
g	6	1	68.2	69.6	0.2
h	6	1	−17.4	−20.2	−0.4
g	6	2	74.2	76.6	−0.2
h	6	2	63.7	54.7	−1.9
g	6	3	−160.9	−151.1	2.1
h	6	3	65.1	63.7	−0.4
g	6	4	−5.9	−15.0	−2.1
h	6	4	−61.2	−63.4	−0.4
g	6	5	16.9	14.7	−0.4
h	6	5	0.7	0.0	−0.2
g	6	6	−90.4	−86.4	1.3
h	6	6	43.8	50.3	0.9

Let us introduce a new point (θ_0, λ_0) and the corresponding value of horizontal intensity H_0, whose meaning will be made clear in what follows, and assume the following identities

$$g_1^0 = H_0 \cos\theta_0 \; ; \quad \frac{h_1^1}{g_1^1} = \tan\lambda_0 = \frac{k\sin\lambda_0}{k\cos\lambda_0}$$

$$g_1^{0^2} + h_1^{1^2} + g_1^{1^2} = H_0^2$$

Then we have

$$H_0^2 = (H_0^2 \cos^2\theta_0 + k^2 \cos^2\lambda_0 + k^2 \sin^2\lambda_0)$$

$$H_0^2 = (H_0^2 \cos^2\theta_0 + k^2)$$

$$H_0^2(1 - \cos^2\theta_0{}^2) = k^2 = H_0^2 \sin^2\vartheta_0$$

$$k = H_0 \sin\theta_0$$

therefore

$$g_1^1 = H_0 \sin\theta_0 \cos\lambda_0 \; ; \quad h_1^1 = H_0 \sin\theta_0 \sin\lambda_0 \; ; \quad g_1^0 = H_0 \cos\theta_0$$

We can now obtain for the magnetic potential V_1 the following expression obtained by the insertion of the Gauss coefficients new formulation

$$V_1 = \frac{a^3}{r^2}\left[H_0\left(\cos\theta_0 \cos\theta + \sin\theta_0 \sin\theta \cos\lambda_0 \cos\lambda + \sin\theta_0 \sin\theta \sin\lambda_0 \sin\lambda\right)\right]$$

$$V_1 = \frac{a^3}{r^2}\left[H_0\left\{\cos\theta_0 \cos\theta + \sin\theta_0 \sin\theta\left(\cos\lambda_0 \cos\lambda + \sin\lambda_0 \sin\lambda\right)\right\}\right]$$

$$V_1 = \frac{a^3}{r^2}\left[H_0\left\{\cos\theta_0 \cos\theta + \sin\theta_0 \sin\theta \cos(\lambda - \lambda_0)\right\}\right] \tag{1.35}$$

where the values of H_0, θ_0 and λ_0 can be obtained from the Gauss coefficients using the relationships previously introduced. Once g_0^1; g_1^1; h_1^1 coefficients are known the relation for V_1 can be written as follows

$$V_1 = \frac{a^3 H_0}{r^2}\cos\Theta = \frac{M}{r^2}\cos\Theta \tag{1.36}$$

where the new angle Θ, can be obtained from the spherical trigonometry cosine theorem[2]. This is the angle between a new pole on the Earth's surface, whose coordinates are (θ_0, λ_0) and that we will name *geomagnetic pole*, and the point of observation $P = P(\theta, \lambda)$.

The expression obtained for V_1 is that of the potential of a magnetic dipole placed at the center of the Earth, whose axis intersects the Earth's surface at the point of coordinates θ_0 and λ_0 so that the colatitude from point $P(\theta, \lambda)$ referred to the new dipole axis, will be given by Θ. H_0 represents the horizontal component of the magnetic dipole field on the Earth's surface in the dipole equatorial plane (Fig. 1.13). On the Earth, being g_1^0 negative the geomagnetic north pole in the northern hemisphere corresponds to a magnetic south pole for the dipole at the Earth's center. Denoting by ϕ_0 the latitude $(90° - \theta_0)$, we can obtain the location of the geomagnetic poles

$$\phi_0 = \arctan\left(g_1^0 / \sqrt{g_1^{1^2} + h_1^{1^2}}\right); \quad \lambda_0 = \arctan(h_1^1 / g_1^1) \qquad (1.37)$$

In the year 2000 the geomagnetic North Pole was located at $\phi_0 = 79.542°$ N, $\lambda_0 = 71.572°$ W.

The potential V expansion can be extended to include also all terms for $n = 2$. These terms are called quadrupole terms. Kelvin showed that the sum of all terms for $n = 1$ to $n = 2$ included, gives rise to a magnetic field similar to that generated by a magnetic dipole parallel to the centered dipole, but displaced with respect to the Earth's center by about 500 km in the direction of the West Pacific Ocean plus a term of minor importance in P_2^2. This new dipolar representation of the Earth's magnetic field is that obtained with an *eccentric dipole* and of course it gives a better fit to the global representation of the Earth's magnetic field than with the *central dipole* obtained only by using terms for $n = 1$.

1.2.5
Geomagnetic Coordinates

Geomagnetic coordinates are defined using colatitudes and longitudes in the frame of the geomagnetic dipole, obtained for $n = 1$. Colatitudes Θ, are angles defined with respect to the axis of the geomagnetic dipole, instead of to the usual geographic representation that refers colatitudes to the Earth's rotation axis. Similarly a geomagnetic longitude can be defined with respect to a new zero meridian line, that will be defined in what follows. As in the case of geographic coordinates, the geomagnetic coordinates enable the identification of the position of points on the Earth's surface with respect to a geomagnetic frame of reference. As mentioned above in this new frame it is moreover possible to identify north and south geomagnetic poles as the points, on the Earth's

---

[2] In spherical trigonometry the cosine theorem states that: The cosine of an angle at the center is given by the product of the cosines of the other two angles at the center plus the product of their sines times the cosine of the angle at the surface opposite to the angle at the center.

surface, where the axis of ideal central dipole intersects the surface. Similarly it is possible to define an ideal line on the Earth's surface representing the intersection of the plane passing through the Earth's center orthogonal to the central dipole. This line is called by analogy, the geomagnetic equator.

From the geographic coordinates θ, λ we can get the geomagnetic colatitude Θ of any given point on the Earth surface, as follows

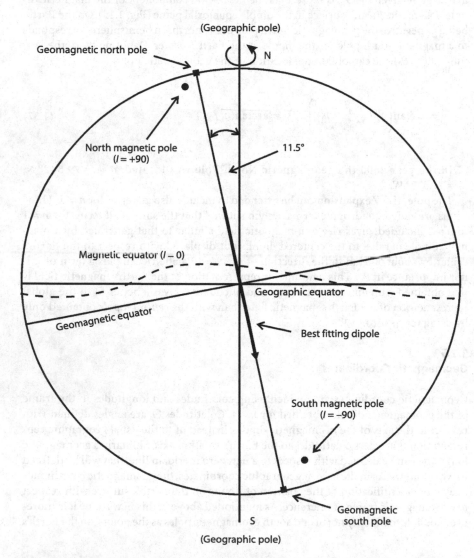

Fig. 1.13. Centered magnetic dipole at the center of a circle representing the Earth with corresponding geographic and geomagnetic poles. On the Earth's surface the geomagnetic poles are only an ideal entity corresponding to the best fitting dipole ($n = 1$). The north and south magnetic poles are measurable points (areas) where dip is $I = \pm90°$. In analogy with the geographic equator a geomagnetic equator can be drawn for geomagnetic latitude 0°

$$\cos\Theta = \cos\theta\cos\theta_0 + \sin\theta\sin\theta_0\cos(\lambda - \lambda_0) \qquad (1.38)$$

which follows directly from the cosine theorem in spherical trigonometry. To obtain the geomagnetic longitude Λ, we firstly define the new zero meridian as the great circle that joins the two poles, geographic and geomagnetic, and then apply the cosine theorem to the θ angle (Fig. 1.14) (geomagnetic longitudes are positive eastward):

$$\cos\theta = \cos\Theta\cos\theta_0 + \sin\Theta\sin\theta_0\cos(\pi - \Lambda) = \cos\Theta\cos\theta_0 - \sin\Theta\sin\theta_0\cos\Lambda$$

$$\cos\Lambda = \frac{\cos\Theta\cos\theta_0 - \cos\theta}{\sin\theta_0\sin\Theta} \qquad (1.39)$$

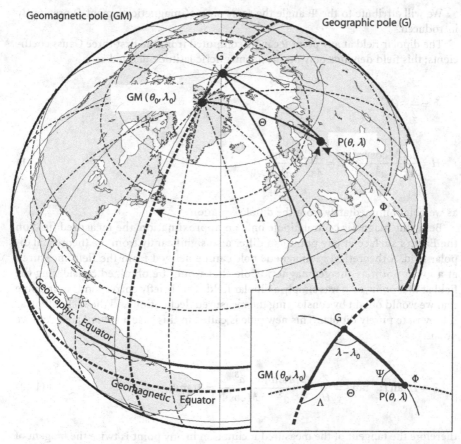

Fig. 1.14. Geographic and geomagnetic coordinates reported on the Earth's surface. *G* geographic north pole; *GM* geomagnetic north pole; *P* is a generic point on the Earth's surface; its geomagnetic coordinates are: Θ = geomagnetic colatitude; Φ = geomagnetic latitude; Λ = geomagnetic longitude

By the sine theorem we can complete the information by obtaining the quadrant in which Λ falls:

$$\frac{\sin(\pi - \Lambda)}{\sin\theta} = \frac{\sin(\lambda - \lambda_0)}{\sin\Theta}; \quad \sin\Lambda = \sin(\lambda - \lambda_0)\frac{\sin\theta}{\sin\Theta}$$

At all points on the Earth's surface, we can also deduce a geomagnetic declination Ψ, as the angle in the new geomagnetic system between the two directions (a) towards the geographic north and (b) towards the geomagnetic north:

$$\frac{\sin\Psi}{\sin\theta_0} = \frac{\sin(\lambda - \lambda_0)}{\sin\Theta}; \quad \sin\Psi = \frac{\sin(\lambda - \lambda_0)\sin\theta_0}{\sin\Theta} \tag{1.40}$$

We will attribute to the Ψ angle the same sign of magnetic element D previously introduced.

The dipolar field at any point P can be computed from the first three Gauss coefficients; this field depends only on the geomagnetic latitude defined as

$$\Phi = \frac{\pi}{2} - \Theta$$

$$H = \frac{\mu_0}{4\pi}\frac{M\cos\Phi}{a^3}; \quad Z = \frac{\mu_0}{4\pi}\frac{2M\sin\Phi}{a^3} \tag{1.41}$$

as we obtain immediately from the dipole equation.

Being the magnetic central dipole only an approximation, the measured field on the Earth's surface at any point can differ also significantly from the theoretical dipolar field. A theoretical geomagnetic pole can be defined from the field measured at a given point as the geomagnetic pole that would be obtained considering the field at that point, as it was simply a dipolar field. This briefly corresponds to the pole that we would obtain by considering the measured declination and the inclination, as if they were purely dipolar. This new pole is called in this case a *virtual pole*. And we obtain

$$Z = H\tan I; \quad \tan I = \frac{Z}{H} = \frac{2M\sin\Phi}{a^3}\frac{a^3}{M\cos\Phi} = 2\tan\Phi \tag{1.42}$$

therefore the tangent of the measured inclination in any point is twice the tangent of the virtual geomagnetic latitude at the point Φ_v.

We must remember that calling φ_{0v} and λ_{0v} the virtual geomagnetic pole coordinates:

$$\cos\theta_{0v} = \cos\Theta_v \cos\theta + \sin\Theta_v \sin\theta\cos D \qquad (1.43)$$

And considering the complementary angles, we will have

$$\sin\varphi_{0v} = \sin\Phi_v \sin\varphi + \cos\Phi_v \cos\varphi\cos D \qquad (1.44)$$

$$\sin(\lambda_{0v} - \lambda) = \frac{\cos\Phi_v \sin D}{\cos\varphi_{0v}} \qquad (1.45)$$

Equation 1.44 follows directly from the cosine theorem, Eq. 1.45 follows directly from the sine theorem. Hence it is possible to obtain the geomagnetic pole coordinates φ_{0v} and λ_{0v}.

1.2.6
Harmonic Power Spectra of the Geomagnetic Field

The classical technique of SHA depicts the spatial structure of the Earth's magnetic field potential in terms of spherical harmonics. The solution is represented by mathematical functions of:

a radial distance from the center of a spherical Earth (considered the origin of a spherical reference system);
b Legendre functions in (cosine of) colatitude;
c Fourier series in longitude.

As a consequence of this operation the resulting basis functions, i.e. the spherical harmonics, are orthogonal over the sphere, very helpful in the practical application of SHA to experimental data. As we have seen before, an important aspect of SHA is that it allows the separation of internal and external contributions to the Earth's magnetic field, although it cannot give the exact location of magnetic sources. These can be imagined as sources together ideally grouped at the origin, i.e. at the Earth's center, for internal sources, and at infinite distance, for external sources. Sources are then given in the form of so-called multipoles, each characterized by a given spherical harmonic degree n.

In the actual case of only internal contributions for degree n and order m up to a given value N_{max}, a global model in ordinary spherical harmonics, can represent details of the field with minimum wavelengths given by

$$\lambda_{min} = \frac{4\pi r}{N_{max}} \qquad (1.46)$$

The possibility of accurate determination of Gauss coefficients for any global magnetic data set is inversely related to N_{max}, therefore SHA is most suitable to model the longest-wavelength part of the geomagnetic field. This is confirmed also by the introduction of the so-called IGRF (International Geomagnetic Reference Field), which is the accepted global model of the geomagnetic field based on spherical harmonics. It is believed to contain all, or most of, the core field, i.e. the largest part of the field observed at the Earth's surface (for this reason also called the main field). The IGRF is given in the form of sets of spherical harmonic coefficients up to degree and order 10. It also includes a separate set of so-called predictive secular-variation model coefficients which extends to degree and order 8. The current 10th generation was updated in the year 2005 and is composed of sets of main-field models, ranging from 1900 to 2005, at 5-year intervals designated as definitive from 1945 to 2000, inclusive. At this time a provisional main-field set for 2005, and a secular-variation predictive model for the interval 2005 to 2010 are the latest available coefficients.

When only internal sources are considered all contributions to the geomagnetic field, i.e. the sum of all Gauss coefficients up to a given degree n, can be shown on in a semi-logarithmic scale plot as a function of the order number n. In Fig. 1.15 the so-called geomagnetic field power spectrum showing R_n vs. n is reported. R_n is expressed in terms of the Gauss coefficients as follows

$$R_n = (n+1) \sum_{m=0}^{n} \left[(g_n^m)^2 + (h_n^m)^2 \right] \tag{1.47}$$

This expression represents the energy contribution brought by every degree n term in the expansion from $n = 1$ to $n = m = 23$. As is clearly observable, and understandable from the Gauss coefficient values, the contribution to the power decreases as the n order number increases. An isolated point is obtained in the plot for $n = 1$, this corresponds to the centered Earth dipole as shown by the meaning of the $n = 1$ spherical

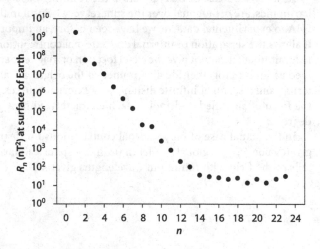

Fig. 1.15. Semi-logarithmic geomagnetic field power spectrum R (nT^2) at Earth's surface for degree $n = 1$ to 23

harmonic analysis. Using the general rule given above, Eq. 1.46, we should attribute a maximum wavelength to the magnetic dipole source. All other points in the plot can be easily interpolated by two well separated straight lines. The contributions from $n = 2$ to $n = 12$, 13 decrease to an almost white noise level. From $n = 13$, 14 onwards, a second almost horizontal line can be fitted. All points lying on the first straight line correspond to the maximum intensity values of the power spectrum (excluding the dipole point) and are related to the contribution of the deepest sources, reasonably located in the fluid core up to the core-mantle boundary. The second straight line is almost horizontal instead and corresponds, in the first approximation, to the crustal sources (possibly with a core contamination) with largest n values almost corresponding to a white noise. For this reason this power spectrum plot cannot express the crustal contribution separately from the white noise.

Magnetic satellite data can come in the picture to give better models of core sources and also help to discriminate the largest harmonic degree n values from the white noise. In fact not only Earth's surface data but also satellite data can be used to compute Gauss coefficients. As will be shown in other parts of the text only a few satellites were in fact used to investigate the Earth's magnetic field over about the last thirty years. In Fig. 1.16 a power spectrum plot from satellite data is shown. In this case a possible crustal field contribution can be inferred for the part of the spectrum that goes from $n = 14$ onwards, and that shows up clearly as a different line with a different slope from that for $n < 14$.

1.3
Time Variations

Time variations of the Earth's magnetic field can be divided into two classes: those having an internal origin and those having an external origin with respect to the Earth's surface. Although it is not precisely possible to fix a clear limit between the two classes, by applying the spherical harmonic analysis to the Earth's magnetic field time varia-

Fig. 1.16. Semi-logarithmic geomagnetic field power spectrum R (nT^2) at the satellite altitude for degree $n = 1$ to 60. Interpolating lines indicate best fit lines for the two emerging preferred slopes

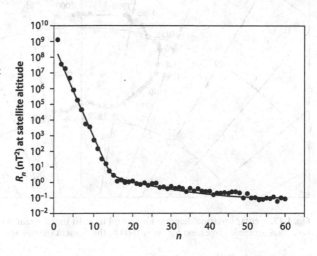

tions, it is shown that variations on a time scale shorter than 5 years are generally considered to be of external origin. The variations on time scales longer than 5 years are commonly called *secular variation* (SV) and are of internal origin to the Earth. External origin time variations are clearly recognizable in magnetic observatory data, survey time variation recording stations, and data from rapid run magnetic time recording devices. Very long period external origin time variations, such as for example, those related to the solar cycle (about 11 yr), can generally be seen only in observatory data. Internal origin time variations, like SV, can be seen in observatory data but also in the archaeological or geological records when magnetic investigations are undertaken on dated archaeological samples or rocks. The amplitude of Earth's magnetic field SV for a given place of observation fluctuates between a few nT/year to several tens of nT/year for the magnetic intensity components and from a few minutes/year to several minutes/year for declination and inclination.

1.3.1
Secular Variation

Secular variation is clearly seen in geomagnetic observatory data, when several years for one or more field elements (generally by their annual or monthly means) are plotted against time. Starting from magnetic field observations, carried out for declina-

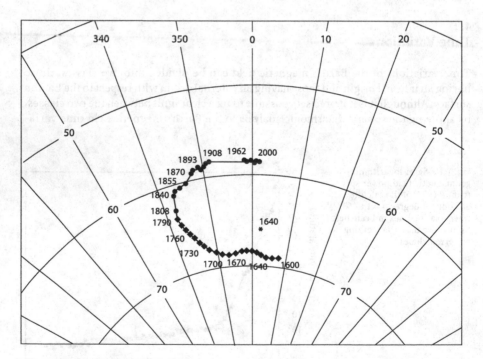

Fig. 1.17. Secular variation diagram for central Italy. In the diagram *D* and *I* time variation is reported on a stereographic projection for years 1600 to the present (from Lanza et al. 2005)

tion and inclination since the 16th century, the geomagnetic field has undergone a SV that, for instance in declination, has covered, in central Europe (Fig. 1.17), a maximum range of about 20°. The angle of declination, in the period under examination, has been predominantly negative, i.e. westerly. SV, thought to be closely linked to the dynamics of the Earth's core and to the phenomena that produce the field itself, occurs on two timescales which are related to two types of core process. One, evident on time scales of hundreds or thousands years, is related to the main dipole field variation, while the second one, clearly appreciable also on the shortest time scales, of the order of tens of years, is related to the non-dipole field variations. For this reason, even though SV shows a different behavior in a range of world observatories, it is typical of the main field, thus being representative of planetary phenomena.

1.3.1.1
Dipole Field Variation

The characteristics of the Earth's dipole field were determined for the first time with modern scientific accuracy by Gauss, by means of his spherical harmonic analysis in 1839, which subsequently allowed to infer the characteristics of the global variations of the field over the last 160 years. In particular the dipole magnetic moment has decreased from 9.6×10^{22} to 7.8×10^{22} A m² from the time of Gauss to 2000 (Fig. 1.18), while the angle between the dipole axis and the Earth's rotation axis has remained almost unchanged at about 11.5°. Moreover the dipole has shown a precessional motion around the rotation axis taking the north geomagnetic pole from its position of 63.5° W in 1830 to 71.6° W in 1990 with a precessional velocity of about 0.05° yr⁻¹. A third time variation of the dipole field consists in the slight displacement of the dipole along its axis towards the geographic north at a velocity of about 2 km yr⁻¹. Inspection of the data from 1600 onwards has shown that an ideal line of force of the purely dipole field has moved over the last 400 years in a westerly direction at a velocity of about 0.08° yr⁻¹ and with a variation in latitude of about 0.01° yr⁻¹.

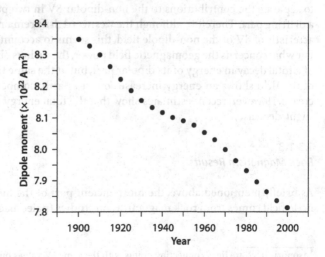

Fig. 1.18. Geomagnetic dipole moment time variation from years 1900 to 2000

The determination of the elements of the Earth's field before instrumental measurements are available, i.e. before roughly 1600, can be done with other techniques. By means of rock magnetism and artefacts' magnetic properties, attempts have been made to determine the value of the dipole moment in the past. For the last thousands years, in spite of all the uncertainties in measuring the paleointensity of the magnetic field, there is sound evidence that the present decrease of the magnetic moment began about 2 000 years ago when its value was around 11×10^{22} A m^2. It has been attempted to establish if the Earth's field has had in the past the same dipolar nature as it has today. It has been inferred that the field is due to an almost axial dipole at least for the last 500 000 years. But this axiality is only of a statistical nature as there are evident and repeated fluctuations of the geomagnetic poles around the geographic poles; the present situation must be considered as one of these fluctuations.

1.3.1.2
Variation of the Non-Dipole Field

Observing contour maps of declination for different epochs, geomagnetists have noted a clear drift of almost all declination contour lines toward the west. Halley was the first to show declination contours on his famous Atlantic Ocean map at the very beginning of the 18th century. This phenomenon is known as *westward drift* and can still today be clearly seen in the Atlantic and Europe, while it is not quite appreciable in the eastern Pacific, Australia and Antarctica. The westward drift, that in many studies on SV has been considered as its most evident characteristic, is due to the variation of the non-dipolar part of the field. Bullard was the first to estimate that the isoporic foci[3] of the non-dipole field undergo, in those regions where the phenomenon is evident, a westward drift motion of about $0.2° \text{yr}^{-1}$. More in-depth studies have shown the possibility to discriminate two different behaviors of the non-dipolar magnetic field contribution to SV: a clear westward drift in some zones, as for example the so-called "African anomaly" (Fig. 1.19), and a geographically stationary effect with a strong intensity variation, for instance the "Mongolian anomaly". This circumstance has suggested to separate the contributions to the non-dipolar SV in two parts: a standing part and a drifting part. Therefore although the westward drift remains the most evident characteristic of SV of the non-dipole field, this seems to account for only one of its parts. For what concerns the geomagnetic field power, the various observations give evidence of a total decay in energy of its dipolar part, but at the same time the non-dipolar part of the field shows an energy increase so as to partly compensate for the dipolar decrease. However, recent estimates show that the total energy probably shows overall a slight decrease.

1.3.1.3
Rock Magnetism Results

As briefly mentioned above, the most ancient part of the history of the Earth's magnetic field comes from rock magnetization studies (paleomagnetism). Accurate analy-

---
[3] Isoporic lines are lines connecting points with the same SV values on a map.

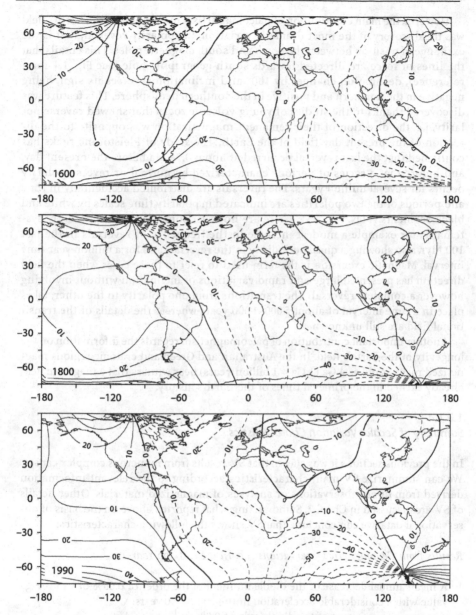

Fig. 1.19. Westward drift of Earth's magnetic field shown by three declination maps at years 1600, 1800 and 1990. Maps made using historical records database and model by Jackson et al. 2000

ses of volcanic rocks and lacustrine sediments has produced curves of paleosecular variation (PSV) extending back as far as hundred thousand years (Chap. 7). The accuracy is obviously lower than that of observatory data, but the whole of PSV data allows statistical analyses useful to outline the behavior of the main field.

However, the main contribution of paleomagnetism to the study of the Earth's field was the discovery of the most dramatic variation, the polarity reversal, that is the exchange of position between the north and south magnetic poles. This entails that the lines of force are directed towards south geographic pole (see Fig. 1.1 for a reference), declination changes by 180° and inclination reverses its sign, being negative in the northern and positive in the southern hemisphere. This feature was discovered thanks to the studies of young volcanic rocks that showed reverse polarity, i.e. the direction of their remanent magnetization was opposite to the direction of the present day field of the Earth. The study of Pleistocene rocks has confirmed that the last reversal occurred at about 0.78 Ma before the present day, and before that a series of periods characterized by normal or reverse polarity occurs for several million years. The reversals are distributed accidentally in time and periods of the two polarities are indicated in polarity time scales by white and black segments (Figs. 7.1, 7.5). There are however some patterns apparently systematic, for example a modest increase in the frequency of reversals in the last 100 Myr, also showing a quasi periodicity in the reversal rates for a million year time interval. Magnetic excursion is the term used to refer to those cases when the field direction has presented large and rapid variations of inclination without involving however a complete reversal. The transitions from one polarity to the other, takes place in a time interval of about 5 000–1 000 years whereas the details of the transitional field are still unknown.

Another important contribution of paleomagnetism regards the information on the longevity of the Earth's field. In the Australian and Greenland cratonic regions magnetized rocks as old as 4 Ga (1 Ga = 1 billion years) bear witness that a magnetic field already existed in the primeval times of the Earth's history.

1.3.1.4
Summary of Secular Variation Characteristics

In the previous sections it was shown that SV results from numerous complex causes. We can summarize SV main characteristics according to the relevant information derived from direct observations or analyses of magnetized materials. Other details of SV are discussed in Chap. 7. Summarizing what reported above on the basis of observational data, we can conclude that SV shows the following characteristics:

Results from Geomagnetic Observations (i.e. in the last 400 Years)

1. A mean annual decrease of the dipole moment of the order of 0.05% of its average value with a considerable acceleration in the last 30–40 years.
2. A westward precession of the dipole axis of $0.08° \, yr^{-1}$.
3. A northward displacement of the dipole of the order of $2 \, km \, yr^{-1}$.
4. A westward drift of the non-dipole field, or of a part of it, of $0.2–0.3° \, yr^{-1}$, associated with a possible but not specified southward drift.
5. An intensity variation (increase or decrease) of the non-dipole field at a mean rate of about $10 \, nT \, yr^{-1}$.

Results from Archaeo- and Paleomagnetic Studies

1. Archaeomagnetic information do not confirm in a clear way the existence of the westward drift and suggest therefore that it is not a permanent characteristic of SV.
2. The field fluctuates in direction and intensity, yet averaged on times greater than 100 kyr it is dipolar, geocentric and aligned with the Earth's rotation axis (GAD hypothesis).
3. The field reverses its polarity at a rate which looks random. The polarity remains constant for time periods of the order of 100 kyr to 1 Myr (Chap. 7).

1.3.2
Magnetic Tomography and Interpretation of Secular Variation

The interpretation of SV has closely followed studies and theories of the Earth's magnetic field generation. The existence and the time variation of the dipole field are attributed to a very deep source in the core, while the non-dipolar variations, in particular the westward drift, are attributed to more superficial sources, probably in relation to electrical currents flowing at the core-mantle boundary (a transition zone that we will denote from now on by CMB). This approach leads to consider that the physical conditions of the lower mantle could influence the dynamics of the core and therefore the geomagnetic field itself. By means of inverse numerical techniques and the spherical harmonic analyses, it is possible to obtain a picture of the magnetic field at the CMB. These maps have recently been produced by several authors and report the field at the CMB where, according to dynamo theory, the electric currents that generate the non-dipole field flow (Sect. 1.3.3). This technique has been named magnetic tomography. One of the most important results obtained by magnetic tomography is that the main contribution to the geometry of the dipole field is given by four patches of intense magnetic flux located symmetrically with respect to the equator, whose positions have not changed significantly from 1600 to nowadays. These four stationary spots have been interpreted as the extreme ends of two columns of fluid material that are tangential to the inner core and extend parallel to the Earth's rotation axis.

In order to get an interpretation of SV, and in particular of westward drift, we need to follow the historical as well as the most recent theories. To explain the presence of the significant westward drift of the field, Bullard in the fifties of the twentieth century, supposed that the core rotated at lower angular velocity with respect to the mantle and proposed as mechanism for this differential rotation the existence of an electromagnetic coupling. Hide in the same period proposed instead the existence of some undulating motions inside the core, able to generate the drift of the field measured on the Earth's surface. The results provided by the magnetic tomography, partly contradict both theories. According to the former the drift should occur everywhere on Earth, while it mainly concerns the region comprised between longitudes from 90° W and 90° E, roughly from Europe and Africa to North and South America. The latter theory, instead, does not explain the absence of SV observed on the Pacific Ocean.

Using magnetic tomographic maps and some approximations from magnetohydrodynamics, it is possible to obtain a configuration of the fluid flow at the CMB. In order to recreate a configuration of the flow, an approximate estimate of the value of the electrical conductivity of the core is necessary. Its mean value is commonly assumed to be $\sigma = 10^5 - 10^4$ S m^{-1} that is $10^3 - 10^4$ times higher than the conductivity estimated for the mantle. The maps of velocity and direction of the flow at the CMB, have shown two fluid circulations below the Atlantic, one toward north and the other toward the south of the equator; near the equator this flow goes in a westerly direction. Therefore the westward drift, observed between longitudes 90° W and 90° E could be the result of this circulation. With the intent of explaining the origin of the fluid flows, in his fundamental fifties works, Bullard was the first to propose a way to generate fluid circulations in the core. Bullard proposed the existence of flows coming from the deep core moving towards the CMB. In this vertical motion, the rising fluid, in the case of frozen field (see Sect. 1.4.2), tends to concentrate the magnetic lines of force against the CMB, so forming a magnetic flux spot alike those in the solar photosphere. Some bipolar magnetic structures form and move quickly, while others remain stationary. For this reason a number of authors have thought that their motion could be modified by some influence external to the core. According to Hide, a possible candidate is the topography of the CMB whose extensions in the core could be envisaged as upturned mountains. Bloxham and Jackson have afterwards excluded this hypothesis as well as that of an electromagnetic coupling. These authors propose that the flow in the core is coupled to the mantle in a thermal way instead. In practice the core fluid would flow upwards below the hot regions of the mantle and downward below the cold ones. Regarding the dipolar part of the Earth's field magnitude the tomographic analysis makes it clear that the present drastic decrease of the dipolar component can be due to the increase and propagation of structures with an opposite flow, below Africa and the Atlantic.

1.3.3
Geomagnetic Jerks

On average magnetic field elements, when plotted versus time at a geomagnetic observatory, show quasi-stable, slow changing time variations. However one peculiar feature of SV is represented by a clear tendency, for a given field element, to show at times rapid changes, observable as a variation in the slope, taking place in one or two years. This peculiar phenomenon that separates periods of reasonably steady SV patterns, i.e. constant slopes in the geomagnetic field time variation, is called a geomagnetic jerk (GJ). GJs are thus abrupt changes in the second time derivative (secular acceleration) of the geomagnetic field. In this sense a GJ separates periods of almost steady secular acceleration of the geomagnetic field.

Secular variation models and the role of the peculiar phenomenon of GJs, following their discovery at the beginning of the eighties of the twentieth century, have widely been investigated. According to the majority of scientists this phenomenon is of internal origin and this was shown clearly by the use of spherical harmonic analyses undertaken on the variation field. Of course the investigation of GJs' importance is mainly in terms of possible explanations of the mechanisms that produce and maintain the geomagnetic SV. We know in fact that SV is associated, via the induction equation in magnetohydrodynamics, to steady fluid flows at the top of the fluid Earth's core;

non-steady, time-varying flows could be associated with SV anomalies, including jerks. The very rapid time variation that is characteristic of GJs is an indication that can also have very important connections to the knowledge of the electrical conductivity of the Earth's mantle. In fact 2–4 years is the accepted upper limit for a magnetic time variation period that is able to to penetrate the full thickness of the mantle without being completely screened out.

Some authors have also associated jerks with decadal fluctuations in the length of day and the Chandler wobble, although this connection is still under examination.

In many of the world magnetic observatory data series, GJs can easily be identified. Jerks are often easier to be observed in D or Y component. For a quantitative study, direct recognizable patterns due to effects of external fields, are generally removed from data series by modeling external variations by means of so-called geomagnetic indices (K, Dst, aa; see Sect. 1.5.3). Last century widely observed geomagnetic jerks are reported in Fig. 1.20 for two observatories, Tucson and Chambon-la-Forêt. Evident slope changes are observed in SV at years 1901, 1925, 1932, 1949, 1969, 1978, 1991 and 1999. The 1969 jerk was discussed in detail in many papers and it was shown that it was observed worldwide although it was not always manifest in all magnetic elements. Other analyses have shown that the occurrence time for GJs, although global in scale, is not the same all over the globe. It is accepted now that for every GJ there is an average time of occurrence and that two to four years are required for the event to be observed all around the globe.

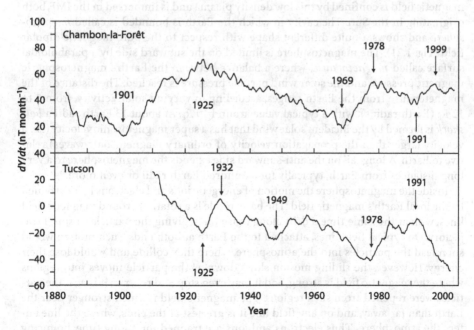

Fig. 1.20. Geomagnetic jerks for element Y the two magnetic observatories Tucson and Chambon-la-Forêt. Slope changes for declination SV are reported with *arrows*

1.3.4
External Origin Time Variations

Time variations of the Earth's magnetic field are constituted by a series of contributions, each of them having a different typology, intensity and typical timescale. As we have just seen, in the case of magnetic field variations, the term *slow* refers to SV; the term *rapid* is referred to external to the Earth origin time variations, mainly connected with the Sun and solar activity. The Sun is in fact the critical factor to interpret all rapid magnetic time variations.

1.3.4.1
Solar Wind and the Magnetosphere

The regular emission of solar electromagnetic radiation is accompanied by a continuous emission of an ionized gas, called the solar wind, which is essentially the expansion of the solar corona. Because of its high temperature, the solar wind leaves the immediate vicinity of the Sun, fills all the interplanetary space and also affects the Earth. The solar wind particle flux is composed mainly by electrons, protons, He nuclei, and also heavier elements and is constantly emitted by the Sun at an average speed of 1.5 million km hr^{-1}; the solar wind density at the Earth is ~7 ions cm^{-3}. The solar wind is not uniform, although always directed away from the Sun, it can change speed and carries a feeble magnetic field of solar origin, the interplanetary magnetic field (IMF). As a consequence of the solar wind expansion in the solar system, the Earth's magnetic field is confined by this low density plasma and is immersed in the IMF, both originating in the Sun. The cavity in which the Earth is bounded is called *magnetosphere* and shows a quite different shape with respect to the pure magnetic dipolar field (Fig. 1.21). The magnetosphere is limited on the sunward side by a paraboloidal surface called *magnetopause,* where a balance between the Earth's magnetospheric magnetic pressure and the solar wind particle pressure is reached. The distance of the magnetopause from the Earth ranges, according to varying solar activity, from 5 to 12 R_e (Earth radius), with a typical value around 10 R_e. At about 13 Earth radii a *bow shock* is formed by the incident solar wind that has a super magnetosonic velocity (i.e. a velocity larger than the propagation velocity of ordinary magnetosonic waves) relative to Earth. A long tail on the anti-sunward side extends the magnetosphere to a very long distances from Earth, typically for a hundred Earth radii or even more.

Inside the magnetosphere the motion of energetic ions and electrons is constrained by the local Earth's magnetic field. The basic mode is a rotation around magnetic field lines, with at the same time a drift along those lines, giving the particles a spiral trajectory. On typical field lines, attached to the Earth at both ends, such motion would soon lead the particles into the atmosphere, where they collide and would lose their energy. However, the sliding motion slows down as the particle moves into regions where the magnetic field is strong, and it may even stop and reverse; it is as if the particles were repelled from such regions. The magnetic field is much stronger near the Earth than far away, and on any field line it is greatest at the ends, where the line enters the atmosphere. Thus electrons and ions are trapped for a long time, bouncing back and forth from one hemisphere to the other. In this way the Earth is encircled by particle radiation belts.

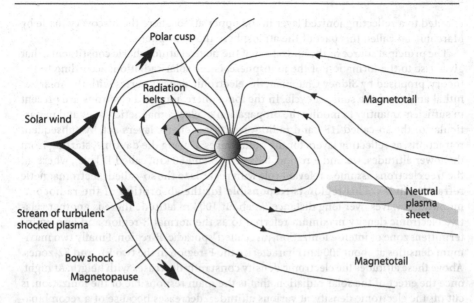

Fig. 1.21. The Earth's magnetosphere. Solar wind shapes the Earth's magnetic field differently from a purely dipolar structure. In the pictorial representation magnetic field lines are drawn to represent the real structure of the Earth's magnetic field around the Earth

From the first satellite investigations it was discovered that the Earth has two regions of trapped fast particles. The inner radiation belt discovered by Van Allen, is relatively compact, extending perhaps one R_e above the equator. It consists of very energetic protons, a by-product of collisions by cosmic ray ions with atoms of the atmosphere. The number of such ions is relatively small, and the inner belt therefore accumulates slowly, but because trapping near Earth is very stable, rather high intensities are reached, even though their build-up may take years. Further out is the large region of the outer radiation belt, containing ions and electrons of much lower energy; this population fluctuates widely, rising when magnetic storms inject fresh particles from the tail then gradually falling off again.

1.3.4.2
The Ionosphere

The magnetosphere has a lower boundary. Theoretically this boundary should be located where the medium passes from the condition of plasma, in which the control of the particle motion is determined mainly by the Earth's magnetic field, to that in which ion and electron densities make collision processes an important factor and the control by the magnetic field small. This layer, that has all the characteristics of a transition zone, can be approximately localized at a height between 200 and 600 km from the ground, to some extent variable with latitude, season, solar time, etc., and is called the ionosphere. The historical discovery of the ionosphere dates back to the early twentieth century, when Guglielmo Marconi succeeded in obtaining an ether connection between Europe and North America. Shortly afterwards, Kennelly and Heaviside at-

tributed to a reflecting ionized layer in the upper atmosphere the discovery made by
Marconi and called this part of the atmosphere the ionosphere.

The principal source of the ionization of the neutral atmospheric constituents, that
gives rise to the formation of the ionosphere, is the solar radiation; according to the
theory, proposed by Sidney Chapman, the electronic density varies with the solar ze-
nithal angle and the sunspot cycle. In the ionosphere, ions and electrons are present
in sufficient quantity to modify the propagation of electromagnetic waves and in particular of the so-called HF band radio waves. Ionospheric layers act, by subsequent
refractions, as reflecting layers for the radio waves. During the daytime, starting from
the lower altitudes, we find a region, between 50 and 80 km, called D layer, where all
the free electrons assume a relevant role with regards to the so-called electromagnetic
refractive index. This region is responsible for the absorption of the radiocom-
munication waves over long distance. At about 100 km altitude there is another relative electronic density maximum referred to as the normal E-region and also an intermittent zone of intense ionization, just called 'sporadic' E-region. Finally two maximum densities at about 300 km characterize the F-region in the two F1- and F2-zones.
Above these altitudes the electronic density constantly decreases with heigth. At night,
once the effect of the solar radiation, that is the main responsible of the ionization, is
cut off the electronic density, at various altitudes, decreases because of a recombination process between ions and electrons; only one peak characterizing the F-region is
in this case observed. Thanks to their improvement, development and the possibility
to use them in various areas of the world, the survey instruments used to measure ionospheric characteristics, the so-called ionosondes, showed that the Chapman theory
could be only applied to two ionospheric layers, E and F1, but not, for example, to the
layer with the maximum electronic density, F2. Therefore other processes had to be
sought to explain why the electronic density of the F2-layer was so high and capable
of resisting also without solar ionization. The diffusive and transport processes taking place in the ionospheric plasma complete the Chapman's photoionization theory.

The presence of the ionosphere and of the ionization processes acting there give
rise to other important geophysical phenomena, besides a radiopropagation by reflec-
tion. As a result of the absorption of the solar radiation, the gas in the upper atmosphere is subjected to an internal photochemical excitation process and also to a natural
ion recombination. This last phenomenon produces a light emission in the infrared,
visible and ultraviolet regions of the spectrum, called *airglow*. Even though it is always present, during the day the airglow cannot be seen easily, because of the strong
sunlight; on the other hand it provides a dim light during the night. The airglow is a
phenomenon that takes place at all latitudes without a particular geometric or temporal structure. On the contrary the aurora (Sect. 1.3.4.4) also caused by photochemical atmospheric processes, takes only place at high latitudes, where energetic particles
of magnetospheric origin act as the ionizing agent.

1.3.4.3
Regular Time Variations

The existence of the ionosphere will help us to explain some of the time variations of
the Earth's magnetic field. The sunlight daily variation is directly connected to the
upper atmosphere electrical conductivity and to the motions of the atmospheric gas

through the Earth's magnetic field lines of force. These motions and their complex interactions with the field, create an electrical current system in the ionosphere detectable on the Earth's surface as a slow modulation of the three components of the Earth's magnetic field, that can be observed clearly only when additional stronger disturbances produced by other phenomena in the magnetosphere, are not taking place. This variation is in fact called Sq from *Solar quiet,* where 'solar' indicates that the variation acts following the local solar time and 'quiet' indicates that it is typical of an unperturbed situation. As the ionospheric conductivity is proportional to the ionic mobility and to ion concentration, the most effective conductive layer is between 90 and 120 km of altitude in the E-region, where the currents responsible for the Sq, are assumed to be located in. The variation, known as *diurnal variation,* acts following the local time; each Earth's magnetic field element shows a time behavior, that can be interpreted as a superimposition of waves with periods of 24 and 12 hours, and their harmonics with an amplitude of the order of some tens of nT. The variation is restricted to the daylight hours; during the night it is negligible. As briefly mentioned on some days it can be clearly seen, but generally it is overlaid by irregular variations that partly deform it, allowing however at times to see its fundamental characteristics.

As it is simpler to study the diurnal variation on days without irregular perturbations, from the very beginning of the twentieth century, it has been agreed to determine, by appropriate methods, for each month, the five most quiet days, valid all over the world (international quiet days) and to calculate the mean diurnal variation during these five days for each observatory. A spherical harmonic study has shown that 2/3 of the daily variation is of external origin to the Earth and can be attributed to electrical currents circulating in the ionosphere and partly in the magnetosphere. In the ionosphere on the daytime side of the Earth these currents are constituted by a couple of fixed vortices one in each hemisphere whose position is fixed with respect to the Sun and thus follows its apparent rotation. The centers of these vortices are in the hemisphere illuminated by the Sun at latitudes of around 40° and very near to the meridian of the Sun. The ionosphere is the dynamo conductor and the Earth's field the magnet. In the nighttime hemisphere there are two vortices of opposite rotation to diurnal ones and of weaker intensity. The remaining 1/3 of the variation is of internal origin and is due to electrical currents produced by electromagnetic induction in the Earth's interior probably extending down to a depth of about 800 km, by the varying external magnetic field. The Sq amplitude changes during the seasons with a summer maximum and a winter minimum at high and mid latitudes, and with a maximum at the equinoxes in the inter-tropical zone for H and Z (Fig. 1.22). Moreover Sq amplitude depends on the phase of the sunspot cycle with the quietest levels occurring around minimum sunspot number years. The ratio between maximum and minimum amplitudes can reach 2 to 2.5. The actual value varies according to the solar cycle intensity as expressed by the sunspot number and the ratio between the amplitude at its maximum and that at its minimum is of the order of 1.5.

Above the equator there is an accumulation of the atmospheric tide in the west-east direction, the so-called equatorial electrojet (EEJ), that causes an enhancement of the diurnal variation in the geomagnetic field near the magnetic equator. This leads to a diurnal variation that can reach values of 200 nT. The width of the interested region is ≈4° in latitude, return currents north and south of the eastward current are a common feature of the EEJ. The intensity of the EEJ varies from day to day.

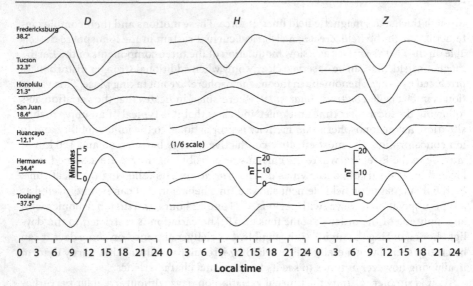

Fig. 1.22. Latitudinal variation of *Sq* for the three magnetic elements *D*, *H* and *Z* as observed on average at the indicated magnetic observatories (from Campbell 2003)

An in-depth study of the daily variation morphology has shown the existence of a weak component which is a function of the lunar time rather than solar. This variation, called *lunar variation L* has a period of 12 lunar hours (12.408 solar hours) and an amplitude of about 1–2 nT. The system of electrical currents generating these regular variations is triggered and sustained by a dynamo process. The atmospheric tides give rise to a wind system in the ionosphere and the motions of electrically conducting matter interacting with the Earth's magnetic field produce in turn electrical currents that are more intensive the higher is the ionization degree, caused by the photoionization in the ionosphere. The atmospheric tides due to the Sun are mainly of thermal origin, those due to the Moon are simply gravitational and for this reason the solar effect is larger than the lunar effect.

1.3.4.4
Irregular Time Variations

In the study of the Earth's magnetic field, in analogy with the weather conditions, the term magnetic storm is used to indicate a generally perturbed magnetic state, characterized by impulses and unpredictable rapid time variations. This irregular phenomenology is strongly influenced by the solar wind. When a magnetic storm starts with a clear impulse (also called a sudden storm commencement, SSC) a typical behavior of magnetic storm time variation evolution is observable globally in the magnetic elements (Fig. 1.23). For example in the initial phase the intensity of the horizontal component increases for a few hours with respect to the average level it had before the storm. Then it decreases quickly down to a lower level than average. This second phase is called the main storm phase and can last from several hours to one or two days. At

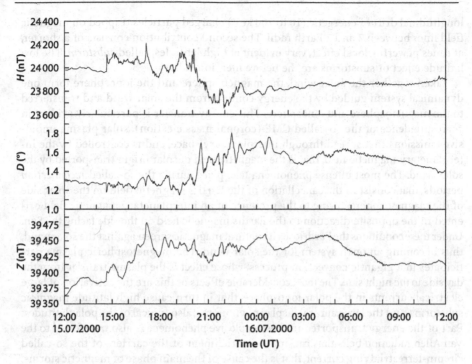

Fig. 1.23. Magnetic elements H, D and Z time variation in the magnetic storm (Bastille event) of July 15, 2000 at L'Aquila Italy geomagnetic latitude 36.3° N

mid latitudes this deviation reaches easily several tens of nT but can reach also hundreds of nT. Finally there is a gradual return to the average value, called recovery phase and lasting generally some hours to one day. The mean total duration of a SSC fluctuates between several hours and a few days. At mid latitudes during the initial and the main phases we can observe also an important activity due to the superimposition of a few more or less intense elementary disturbances, called *bays*. They generally take place during the nighttime hours and last between 1 and 2 hours. Bays are among the most important irregular variations recorded at the magnetic observatories. The bays often coincide with storms, but some times they are observed also without a global magnetic storm taking place. Accordingly they can be considered as elementary storms (also called substorms). From the observations it has been inferred that the bays are produced by electrical currents flowing in the ionosphere at latitudes between 65° and 70°. These currents are produced in the magnetosphere, flow into the ionosphere and return to the magnetosphere following the lines of force of the magnetic field. All irregular variations show up in intensity and forms depending on latitude.

After a careful analysis it was possible to identify different causes for the two main contributions to the irregular variations. The first that characterizes the planetary magnetic storms, consists of a systematic variation of the field that can be mainly attributed to the so-called *ring current*. This is one of the major current systems in the magnetosphere, it encircles the Earth in the equatorial plane and is generated by the longitudinal drift of energetic (10 to 200 keV) charged particles trapped on magnetic field lines between 2 and 7 Earth radii. The second contribution consists of a shorter, at times powerful, local effect, very evident at high latitudes, called *substorm*. The low latitude effect of substorms are the before mentioned bays.

Taken together the solar wind, the magnetosphere and the ionosphere form one dynamical system guided by the energy coming from the solar wind and transferred to the magnetosphere and ionosphere. This energy transfer is particularly evident in correspondence of the so-called CME (coronal mass ejection), solar plasma impulsive emissions that spread through interplanetary space, and is controlled by the interplanetary magnetic field, that is the magnetic field of solar origin transported by the solar wind. The most intense phenomena take place during the so-called *reconnection* periods, that consist in the cancellation of the Earth's magnetic field on the solar side of the magnetosphere, owing to the presence of an interplanetary magnetic field oriented in the opposite direction to the Earth's magnetic field on the side facing the Sun. Under these conditions the Earth loses its natural magnetic screen against the solar wind, thus becoming an open system for the solar wind. The magnetospheric plasma participates in a gigantic convection process, whose effect is the plasma transfer from the day side to the night side. The most considerable effects of this are the activation of large electrical currents in the polar ionosphere, that in turn cause high latitude magnetic substorms, and the large auroral display phenomena, also typical of the polar latitudes. Part of the energy transported in these explosive phenomena is also transferred to the Van Allen radiation belts thus causing the enrichment of the carriers of the so-called circum-terrestrial ring current, that is the cause of the main phase of magnetic storms.

The peaks of solar activity, characterized by an eleven-year cycle, coincide with conditions of greater magnetospheric activity and the intensification of the effects mentioned above. As many technological systems (satellites, high-voltage power lines, oil pipelines, etc.) are sensitive to conditions of high magnetospheric perturbation, most of applied research in this field is devoted to forecasting the magnetospheric activity, in order to prevent damage, in analogy with the weather forecasting.

1.4
Essentials on the Origin of the Earth's Magnetic Field

Following his spherical harmonic analysis of the Earth's magnetic field, in the 19th century, Gauss had analytically demonstrated that the origin of the field was internal to the Earth. Nevertheless not all the scientific community at that time was entirely convinced. For example, although very slow, the observed temporal variation of the elements of the field did not seem in agreement with other phenomena of internal origin to the Earth. Typical manifestations of internal origin showed on average an apparent temporal invariability and all known internal phenomena showed a high stability. Moreover the most rapid magnetic time variations that were discovered thanks to the introduction of photographic recording of variations of the elements of field, did not perfectly appear in tune with an internal origin. For example, the most rapid time variations showed evident correlations with the solar activity; the recorded magnetograms had a strong diurnal periodicity and showed other indications that seemed to suggest a possible external origin. Therefore the origin of the field remained an enigma and a possible source external to the Earth, even if in contrast with Gauss's results, could still not be entirely ruled out.

The most obvious among the hypotheses able to justify an internal origin, was that of a uniform magnetization of the whole Earth. However this hypothesis was set aside very soon. In fact temperatures measured in mines and other deep excavations, were showing that they were very high inside the Earth. Linearly extrapolating measured gradients showed that approximately at 25 km depth, the temperature is higher than the Curie temperature of almost all known ferromagnetic materials. If the magnetization hypothesis had to be followed, then only a thin outer layer of the whole Earth, with a maximum thickness of 25–30 km, should have been considered. In this case a simple calculation leads to an average magnetization of about 10^4 A m^{-1}, a value well above that of crustal rocks. Moreover a permanent magnetization could explain only a constant magnetic field, that contradicts the well-known variability of the Earth's magnetic field.

At the beginning of the 20th century a new impulse to the search for the origin of the Earth's magnetic field came from seismology. The hypothesis that our planet possesses a fluid core, mainly composed of a high-electrical conductivity material such as iron, revitalized the hypothesis for an internal origin of the field but of a different type: a magnetic field generated by an electrical current system. This theory was based in particular on the possible existence of deep Earth conductive fluid motions able to produce electric currents, which consequently generated a magnetic field. The present magnetic field could not be a remnant of an ancient process, a dynamo continuously operating is necessary. Briefly this dynamo can be assumed to operate through a distortion and amplification of an initial magnetic field due to a magnetohydrodynamic interaction with the motions of the plasma constituting the fluid Earth's core. The field so generated decays with time, owing to ohmic dissipation in the conductor. For its maintenance it is then necessary to hypothesize its continuous regeneration to the detriment of some other forms of energy.

The idea of a self-sustaining magnetic dynamo in the Earth was proposed by Larmor in 1919 but it was not immediately accepted; moreover in 1933 Cowling showed that a field with axial symmetry could not be sustained by means of fluid motions with the same symmetry. Towards the end of the 1950s it was possible to demonstrate the existence of a homogeneous self-sustaining dynamo, clearing the way to one of the most fascinating fields of theoretical geophysics. However it was also clear that the introduction of dynamo theories in planetary and stellar physics for the explanation of magnetic fields in the Earth and the Cosmos, represented an extremely complex subject.

1.4.1
Toroidal and Poloidal Fields

In order to get a better understanding of the mathematical scheme of dynamo theory, it is necessary to introduce several mathematical concepts, starting with: toroidal and poloidal magnetic fields. From Maxwell's equations, for a solenoidal magnetic field (hereafter referred to the magnetic induction B), it is always possible to obtain a potential vector A

$$\nabla \cdot B = 0 \;\Rightarrow\; B = \mathrm{curl}A = \nabla \times A \tag{1.48}$$

where

$$A = Tr + (\nabla S) \times r \tag{1.49}$$

being T and S two scalar functions introduced to express the potential vector A and r the radius vector, we have

$$B_T = \mathrm{curl}(Tr) = \nabla \times (Tr)$$

$$\nabla \times (Tr) = T(\nabla \times r) + (\nabla T) \times r$$

being by definition $\mathrm{curl}r = \nabla \times r = 0$,

$$B_T = (\nabla T) \times r \tag{1.50}$$

which is called the toroidal field; it is always perpendicular to r that is to say it lies on a spherical surface.

As regards the second function under study S, since for any scalar function $(\nabla S) \times r = \nabla \times (S \cdot r)$; therefore we can write the second term in B as follows

$$B_P = \nabla \times (\nabla \times (S \cdot r)) = \mathrm{curl}^2(S \cdot r) \tag{1.51}$$

which is called the poloidal field and that can have a component along r.

Therefore B can always be represented by the sum of a toroidal and a poloidal magnetic field:

$$B = (\nabla T) \times r + \mathrm{curl}^2(S \cdot r) \tag{1.52}$$

The toroidal field has always components perpendicular to r and so it lies on spherical surfaces. It is also called "electrical mode" since it is similar to the distribution that electric currents normally take, for instance, on a sphere. The poloidal field, on the contrary, can have radial components and is also called "magnetic mode" because it is typical, for example, of dipolar magnetic fields. So what we can measure outside the core is only the poloidal part that represents for us the Earth's surface measurable magnetic field. In Fig. 1.24 low degree examples of poloidal and toroidal fields are shown.

1.4.2
Fundamental Equations of Magnetohydrodynamics

The first fundamental equation of magnetohydrodynamics is the equation of motion for an incompressible electric fluid conductor immersed in a magnetic field. This is then a typical Navier-Stokes equation modified for the existence of a magnetic field

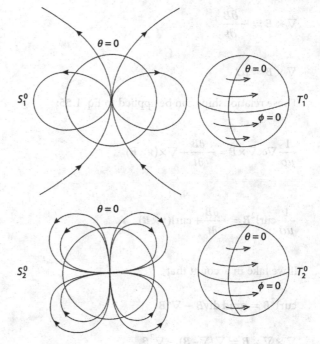

Fig. 1.24. Examples of low degree poloidal (*S*) and toroidal (*T*) field lines on a sphere with θ colatitude and φ longitude (redrawn from Merrill et al. 1996)

$$\frac{d\boldsymbol{v}}{dt} = -\frac{1}{\rho}\nabla p - 2\Omega \times \boldsymbol{v} + \eta\nabla^2\boldsymbol{v} + \Im + \frac{1}{\rho}\boldsymbol{J}\times\boldsymbol{B} \tag{1.53}$$

where p denotes the pressure; Ω is the Earth's rotation; $\Im$ denotes the volume forces; η is the coefficient of kinematic viscosity; ρ is density, $\boldsymbol{v}$ is fluid velocity and $\boldsymbol{J}$ the electrical current density.

To introduce a second fundamental equation we need now to go back to some electrodynamic principles. In general, Ohm's law can be written in a vectorial form as

$$\boldsymbol{J} = \sigma\boldsymbol{E} + \sigma\boldsymbol{v}\times\boldsymbol{B} \tag{1.54}$$

where σ is the fluid electrical conductivity. Consider the curl of both sides in this equation:

$$\frac{\nabla\times\boldsymbol{J}}{\sigma} = \nabla\times\boldsymbol{E} + \nabla\times(\boldsymbol{v}\times\boldsymbol{B}) \tag{1.55}$$

In case it is possible to neglect the contribution of the displacement current, as may reasonably be assumed in a high conductivity environment, from the Maxwell's equations we have

$$\nabla \times E = -\frac{\partial B}{\partial t}$$

$$\nabla \times B = \mu J$$

These relationships can be applied to Eq. 1.55:

$$\frac{1}{\mu\sigma}\nabla\times\nabla\times B = -\frac{\partial B}{\partial t} + \nabla\times(v\times B) \tag{1.56}$$

$$\frac{1}{\mu\sigma}\mathrm{curl}^2 B = -\frac{\partial B}{\partial t} + \mathrm{curl}(v\times B)$$

If we take in account that

$$\mathrm{curl}^2 B = \mathrm{grad\ div} B - \nabla^2 B$$

$$\nabla\times\nabla\times B = \nabla(\nabla\cdot B) - \nabla^2 B$$

we obtain the second fundamental equation of magnetohydrodynamics:

$$\frac{\partial B}{\partial t} = \frac{1}{\mu\sigma}\nabla^2 B + \nabla\times(v\times B) \tag{1.57}$$

According to this equation time variations of the Earth's magnetic field, its possible increase or decrease, are connected with the velocity of the motion of the fluid conductor present in the core. In order to illustrate this situation, we can say that the increase or decrease of the field is due to the conditions of instability represented by the two terms on the right side in Eq. 1.57. The first term is called *diffusive* and expresses the possibility that in a given place the magnetic field decays in a characteristic time; in fact the term $1/\mu\sigma$ is also called magnetic diffusion. The second term, on the contrary, can be called *advective* and expresses the close connection between the field and the motions of the fluid, which can in fact increase the magnetic field to the expense of the fluid's kinetic energy. This situation can be compared with that of meteorological advection which explains that an atmospheric temperature variation, at a given place, occurs for two distinct causes: a local variation, or the transport of warmer/colder air from other zones as a consequence of atmospheric motions. Briefly, the second magnetohydrodynamic equation, also called induction equation, tells us that, inside the Earth's core, a local modification of the magnetic field can be attributed to only two causes: diffusion and advection. Moreover we can simplify this by saying that, in the case of high-electrical conductivity, that we can reasonably assume to be present in the Earth's core, the diffusive term is far lower in importance than the advective term.

Under appropriate conditions of very high-electrical conductivity, the diffusive term in Eq. 1.57 could in the extreme case be neglected. Therefore the temporal variations of the field in this case are only connected with the fluid velocity configuration. In these conditions we obtain the so-called *frozen flux* magnetic field hypothesis. The concept of frozen magnetic field is attributed to the physicist Alfvèn who was the first to show that the magnetic field lines of force in a fluid in motion, in the case of perfect-electrical conductivity, can be considered as frozen in the fluid; therefore changes in magnetic field are completely due to advection of the field owing to fluid flow, where the magnetic field lines act as markers of fluid motion. Even if under different physical conditions, this situation has been found to apply also in the Sun and in the solar wind, where the motion of plasma transports the magnetic field of solar origin and leads it to fill the interplanetary space.

When a magnetic field is immersed in a fluid conductor, the magnetic field can be transported and deformed by the fluid motion, under appropriate conditions of dimensional factors, velocity and fluid electrical conductivity. It can also be said that, in a perfect conductor, the magnetic field flux out of a surface in motion with the fluid, is constant. This concept can also be visualized with the magnetic field lines of force. If the ideal surface in motion is deformed, the magnetic field lines can be used as tracers of the fluid motion.

The fundamental advantage of the introduction of the induction equation in dynamo theory is that it reduces the relevant quantities, such as all the dynamo problem variables, to only two vector quantities: the magnetic field and the fluid velocity or, to be more precise, the fluid velocity field. On the whole, four Maxwell equations and Ohm's law are replaced by only one equation. If we include in the analysis the equation of motion, we have two equations for two unknown quantities: the magnetic field and the fluid velocity field. However in order to eliminate the diffusive term in the induction equation, we have supposed that the fluid in the core has an infinite conductivity. This condition is called *magnetohydrodynamic condition*. Obviously, being this an unrealistic condition, it must be replaced by an extremely high but not infinite conductivity. An important consequence of the introduction of this restriction will be a slight attenuation of the field with time, that is a non-zero diffusive term, that must be balanced by a continuous regeneration of the field through the dynamo process.

1.4.3
Elementary Dynamo Models

A magnetic field can be generated if an interaction between the fluid motion and the magnetic field takes place, consequently an exchange of energy between the fluid kinetic energy and the magnetic field can take place. If the fluid motion geometry shows a velocity gradient, under the hypothesis of the magnetohydrodynamic condition, this leads to a concentration of the magnetic field lines and therefore to an increase in magnetic field intensity. We can observe that in this case the magnetic field energy has increased (larger density of field lines) where the velocity gradient is present to the expense of the fluid kinetic energy (Fig. 1.25). Several problems arise in dynamo

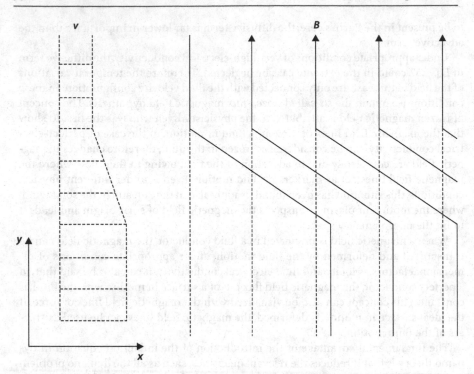

Fig. 1.25. Fluid velocity v gradients in a plane (x-, y-axes) and corresponding B magnetic field lines concentration leading to a field increase (redrawn from Merrill et al. 1996)

theories, we will here only mention a few of them. One is the origin of the energy necessary for the dynamo to work, another is to demonstrate if a certain configuration of motion in the fluid inside the core is able to generate the field as measured on the Earth's surface. This second "preliminary" problem is called that of the kinematic dynamo and has been already discussed since the 1950s; so far many theories had proved the possibility to produce magnetic fields with certain fluid motions. Some empirical numbers are used in the description of the complex dynamics of fluid motions. The Reynolds number is an example. The Reynolds number is defined in terms of some geometrical parameters of the fluid and, according to its value, the fluid motion results to be laminar or turbulent. In the case of a fluid immersed in a magnetic field the magnetic viscosity

$$\nu_\mathrm{m} = \frac{1}{\mu\sigma}$$

is used to define the so-called *magnetic Reynolds number*

$$R_\mathrm{m} = \frac{L \cdot V}{\nu_\mathrm{m}}$$

where L and V denote a geometric scale and a typical velocity of the fluid motion. In short, this number expresses the relationship between the geometric details of fluid motion and its velocity.

In the frame of the dynamo theory, motions that are thought to produce a dynamo effect are also thought to be turbulent. The turbulence takes place when, in a given point of the fluid, the velocity of one of its elements fluctuates randomly without showing a correlation with the velocity of other parts in the fluid, therefore in contrast to the typical laminar flow of slow and organized motions. Turbulent motions are characterized by particles, animated by rotational vortical motions, which create vortices. Fluid vortices have high stability and move in the fluid as if they were individual particles. The value of Reynolds number fixes a quantitative limit between the laminar and vortical regimes. Once the vortical regime has taken place, the presence of a close correlation between its velocity and the turbulence, mathematically expressed by $(\nabla \times V)$ inside the fluid, leads to another important dynamic effect, the helicity, i.e. $V \cdot (\nabla \times V) \neq 0$, which is a specific example of turbulence. The existence of helicity is the manifestation of a strong transfer of kinetic energy from fluid volume motions to fluid vortices. Another important factor in the development of a *turbulent dynamo* was the introduction of the electrodynamic *mean field* approximation. An α-dynamo is a dynamo based on the transformation of turbulent energy of the fluid in electrical energy. In an α-dynamo the electromotive force created by means of turbulent motion is parallel to the magnetic field. This means that the generation of electromotive force due to turbulence will be such that

$$E = \alpha \overline{B}_0 \qquad (1.58)$$

where $\overline{B}_0$ is the initial stationary magnetic field and α an appropriate constant. Only in the case of suitable geometries, this electromotive force can produce an electrical current that strengthens the initial field. The α-effect is fundamental for two of the most well known candidates for terrestrial possible dynamos called simply the α^2 and $\alpha\omega$ dynamos.

We will analyze the $\alpha\omega$-dynamo only in a schematic rather than a mathematical way. Consider an initial poloidal magnetic field S_1^0 and a toroidal velocity field T_1^0 to be present in the Earth's fluid core. In the frozen magnetic flux approximation, the fluid motion deforms the poloidal magnetic field, therefore the pre-existing magnetic field becomes more intense where, due to differential rotation, the fluid flow compresses the poloidal magnetic field lines to the detriment of the fluid's kinetic energy. The interaction produces a new toroidal magnetic field called T_2^0, that has opposite signs in the two hemispheres; this can be schematically expressed as

$$S_1^0(\text{mag}) + T_1^0(\text{vel}) = T_2^0(\text{mag}) \qquad (1.59)$$

This effect is called the ω-effect.

Once the toroidal field has been produced a second fluid velocity field comes in the play. Owing to the reasonable presence of a fully developed convection in the core, a motion with a radial component is now assumed to exist. The convection cells, that can be visualized as columns consisting of fluid substance moving circularly with a volume radial velocity from the core bottom to the CMB and vice versa, will produce, as it happens in the Earth's atmosphere, some ascending rotational motions consisting of fluid substance (equivalent to tropical atmospheric cyclones). This torsion is the effect of Coriolis forces. The net result is a helical motion which interacts with the toroidal magnetic field lines previously generated, that we assume to be normal to the axis of the ascending columns. Original and toroidal field lines will be deformed and undergo a torsion due to helical motion in the column (Fig. 1.26). If electrodynamic forces are such as to limit this torsion to 90°, some circular magnetic coils will form perpendicular to the initial toroidal field line, i.e. in the meridian plane. Finally as a third step new coils will then be able to regenerate magnetic field lines of the initial poloidal field by a *coalescence process*.

Briefly, we have three processes also summarized in Fig. 1.27:

1. Generation of a toroidal magnetic field T_2^0 by the ω-effect (differential rotation)
2. Ascending flows with helicity produce closed coils in meridian planes
3. Closed coils, by a coalescence process, regenerate the S_1^0-magnetic field (α-effect)

The whole described process is intrinsically three-dimensional and can be mathematically explained by means of the differential equations and partial derivatives given in the above mentioned magnetohydrodynamic Eqs. 1.53 and 1.57.

1.4.4
Dynamo Energy

Even if many details of dynamo theory remain unsolved, or are still to be developed, at this time a sufficient schematic knowledge of the physics that determines the process of generation of the terrestrial magnetic field in the core, is available. Electric current systems, produced by a dynamo effect would decay within a time window of about 1 000 –10 000 years unless they were constantly regenerated by an energetic source. Different energy sources have been proposed for the maintenance of the electric current systems flowing in the core and in particular we will take three of them into consideration:

1. A gravitational descent of the heaviest elements from the fluid part of the core into the inner solid part; in this process the inner solid core size increases and a convection mechanism is mechanically generated in the fluid core. This solidification also releases energy due to latent heat of solidification.
2. A thermal convection produced by possible radioactive sources in the core. This effect is however considered less effective than the compositional convection mentioned above, in particular from a mechanical point of view.
3. A coupling between the mantle and the core due to Earth's axis precession (Poincaré had already dealt with this subject), even if many authors think it not to be entirely effective for dynamo from an energetic point of view.

Fig. 1.26. From top to bottom, a magnetic field line, subjected to a fluid upwelling in rotation around a vertical axis in a helical motion (*dashed lines*), is distorted and undergoes a torsion (redrawn from Merrill et al. 1996)

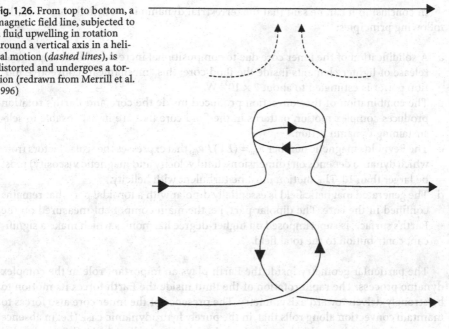

Fig. 1.27. Three processes that lead to the Earth's dynamo action (redrawn from Merrill et al. 1996)

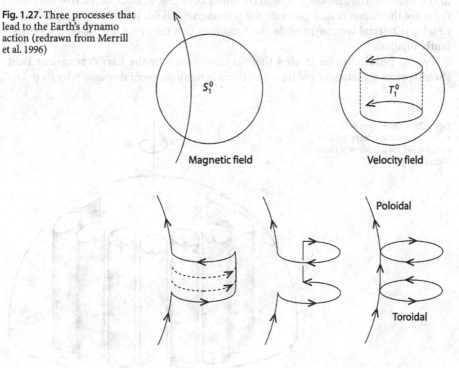

In conclusion it can be said that the terrestrial dynamo is essentially based on the following principles:

a A solidification of the inner core due to compositional increase and corresponding release of lighter elements inside the fluid core; this "mechanical" convective motion power is estimated to about 2×10^{11} W.
b The combination of the convection produced inside the core and Earth's rotation produces complex motion patterns in the fluid core that are indispensable to self-sustaining dynamo action.
c The Reynolds magnetic number $R_m = (L\,V) / v_m$, that expresses the scalar factors from which dynamo depends on (dimensions, fluid velocity and magnetic viscosity) must be larger than 10. The motion must be turbulent with helicity.
d The generated magnetic field is essentially dipolar with a toroidal part that remains confined in the core. The dipolar part, i.e. the main component measured on the Earth's surface, is superimposed on higher-degree harmonics which make a significant contribution to the total field.

The particular geometry inside the Earth plays an important role in the complex dynamo process. The rapid rotation of the fluid inside the Earth forces its motion to be strongly driven toward axisymmetry. The presence of the inner core also forces to maintain convection along rolls that, in the purely hydrodynamic case (i.e. in absence of a magnetic field) would be approximately shaped as cylindrical columns parallel to the rotation axis and tangential to the inner core (Fig. 1.28). Conductive fluid motions are the currents that generate the geomagnetic field. It is likely, because of the nearly cylindrical symmetry of the fluid motions that the geomagnetic field is persistently dipolar.

From a physical point of view there is no reason why the Earth's magnetic field should prefer a particular polarity, and there is no fundamental reason why its polar-

Fig. 1.28. Convection rolls in a rapidly rotating sphere with inner solid core (redrawn from Merrill et al. 1996)

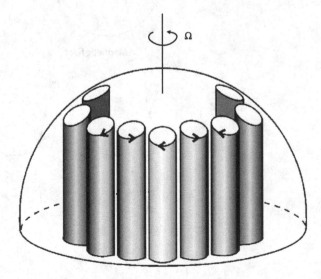

ity should not change. Given the axisymmetry of core fluid dynamics it is possible to see that a dynamo can produce a field in either direction with respect to the rotation axis. A geomagnetic field reversal is in fact an implicit possibility because of the fundamental symmetry of the dynamo equations: if B is a solution, then so is $-B$. Unprecedented details with impressive realizations of geomagnetic field lines and their behavior in terms of secular variation and geomagnetic reversals, are now the results of numerical integrations of fully three dimensional dynamos. The use of supercomputers has allowed a few very specialized groups of scientists to obtain detailed models and plots of the field lines for the Earth's magnetic field.

1.5
Magnetic Observatories, Reference Field Models and Indices

1.5.1
Geomagnetic Observatories

Geomagnetic observatories are the structures designed to undertake a continuous monitoring of the Earth's magnetic field. This is obtained by a standard recording of natural magnetic field time variations and by measuring the absolute level of the magnetic field in all its elements. In geomagnetic observatories this requirement is necessary in order to reveal all possible time variations of interest in geomagnetism, including secular variation, the longest time-scale observable variation. This long-term engagement of an observatory is of course difficult to maintain for several reasons, but when achieved it provides invaluable data for a variety of geophysical studies and in the case of geomagnetism it is the only way to investigate secular variation in detail.

Temporary variometer stations are less restricted structures where magnetic instruments are installed for limited time campaigns in different parts of the world, normally for the recording of rapid external origin time variations. One example is the data acquisition of rapid time variations that need to be known during the execution of magnetic crustal field surveys. Other applications of temporary magnetic stations are for example investigations of the space structure of geomagnetic storms or reconstruction of ionospheric or magnetospheric electrical currents. Conversely in the geomagnetic community, an observatory is considered a 'solid' long-term structure where not only a continuously high accuracy in the absolute level of magnetic field measurements is required, but also a commitment to long time working is necessary.

A continuous long-term monitoring of the Earth's magnetic field is carried out in many observatories all over the world. The number of locations where this is undertaken has, since the times of Gauss, grown to about 150 (Fig. 1.29). Many geomagnetic observatories have at this time collected more than one, or in a few remarkable cases, two centuries of magnetic data. Exceptionally remarkable cases are for example the very long time series of London and Paris, that for the Earth's magnetic field angular elements, go back to the mid-sixteenth century. Plots of magnetic elements versus time, using observatory data, are used for example to study secular variation, the slow unpredictable change of all magnetic elements. From a knowledge of secular variation, several studies have been conducted on the Earth's deep interior and particularly on the electrical conducting fluid motions in the Earth's core.

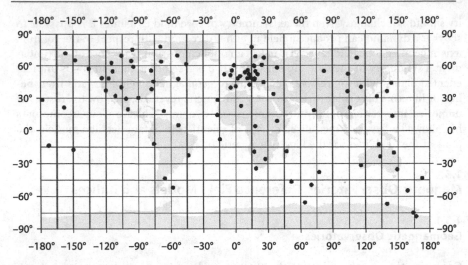

Fig. 1.29. Geomagnetic observatories in the World

From the beginning of magnetic field studies it was recognized that geomagnetic observatories needed to be located reasonably far away from strong natural disturbance such as volcanoes or magnetic mineral deposits, in order to be able to record the average level of the field representing a vast area. Of course, also the need of being away from artificial noise was soon recognized, something that was reasonably easy until only about a century ago. Nowadays the strong perturbation of the natural magnetic field introduced by electrification, especially tramways, underground trains and surface railways, has strongly limited the possibility of installing or even maintaining magnetic observatory activities in many developed countries. This is a problem faced in several geomagnetic institutes. At the beginning, in the few existing magnetic observatories the effort was only to measure the time variations of the angular elements such as D and I, from the mid-eighteenth century also intensive elements were recorded. During the years, magnetic measurement methods have of course changed considerably, especially in the last twenty to thirty years. Only until a few tens of years ago, magnetic operators were engaged with so-called absolute magnetic measurements that could take hours. These reference values for all magnetic field elements were used for the knowledge of the natural base line levels of the natural magnetic field. These procedures are to a great extent simplified today, but a certain skill is still required in the correct conduction of a geomagnetic observatory.

In general, a magnetic observatory is constituted by a few non-magnetic buildings where different tasks are performed. One building is hosting those instruments that run continuously, the variometers, for example fluxgates, in order to record magnetic elements time variations. A second building hosts the so-called absolute measurement devices. These are instruments with which magnetic operators undertake those measurements that are used to calibrate the time variations recorded by the variometers, for example proton precession magnetometers and DI-fluxgates for angular elements. DI-fluxgates are the new magnetic theodolites, consisting of a fluxgate magnetometer element and an optical theodolite. The fluxgate element is mounted axially on an optical theodolite telescope, when the fluxgate element is orthogonal to a magnetic field line, a zero field is measured by the fluxgate electronics. This instrument allows an accurate determination of declination and inclination angles by a null method. Generally in the absolute building a series of pillars host the absolute instruments and one pillar is taken as the fundamental one for the observatory.

Nowadays in order to facilitate data exchanges and the making of geomagnetic products available close to real time, an international coordination programme among all world magnetic observatories called INTERMAGNET is in operation. The programme has facilitated the establishment of a global network of cooperating digital magnetic observatories, adopting modern standard specifications for measuring and recording equipment. An INTERMAGNET Magnetic Observatory (IMO) is a modern magnetic observatory, having full absolute control, that provides one minute magnetic field values measured by a vector magnetometer, and an optional scalar magnetometer, all with a resolution of 0.1 nT. Vector measurements performed by a magnetometer must include the best available baseline reference measurement. Since the effort of running magnetic observatories is elevated, geomagnetic institutions also take magnetic measurements at discrete points, in so-called magnetic networks. These points are marked on maps and are reoccupied every few years (normally at intervals of 5 years), and measurements of all magnetic elements are there repeated. For this reason these points are generally called *repeat stations*. With this additional information geomagnetic observatory data are supplemented and the full information on secular variation behavior on large areas around observatories is obtained. In the case of Italy a network of about 100 repeat stations supplements the magnetic information from the two magnetic observatories, L'Aquila and Castello Tesino.

1.5.2
Geomagnetic Field Reference Models

Data from observatories as well as from repeat stations are only point data. In several occasions the need of magnetic field elements determination in a certain region, or even over the whole Earth, is required. For example the accurate estimate of the magnetic field elements, in particular the magnetic declination, for navigational purposes, is required. For what concerns the portion of the magnetic field generated in the Earth's core, the community of scientists that operates in geomagnetism, has decided to represent it by means of the so-called geomagnetic reference fields. Geomagnetic reference fields are mathematical models of the Earth's magnetic field that represent its space and time variations; time variations generally refer only to secular variation.

Magnetic observatory data, as well as land, marine, airborne and satellite survey field measurements are the fundamental contributions from which models can be made. In conjunction with magnetic observatory data the availability of measurements on a network of repeat stations provides the means to monitor the geomagnetic field and its long-term variations on a regional scale. Geomagnetic observatories and repeat stations are of course located in areas free of artificial disturbances, and not characterized by large surface anomalies and therefore satisfy the standard requirements for reasonably reflecting the main field.

A mathematical model is a representation, for example by means of power series in latitude and longitude, of the spatial time variations of the Earth's magnetic field over a certain area. Mathematical models representing the geomagnetic field at a certain epoch and its secular variation on a global Earth scale, are also possible, generally by means of spherical harmonic polynomials. If secular variation is represented in a certain area, with data from different observatories, for example in a given time interval, a map that shows contour lines of secular variation, can be obtained; these maps are generally called isoporic charts.

One of the most important applications of SV models is obtained when several magnetic surveys are collected in order to join all measurements to a central unifying epoch. Magnetic anomaly maps, for the study of the crustal contribution to the geomagnetic field, are a typical application. Magnetic anomaly maps are in fact constructed after the removal of the main part of the geomagnetic field. Time reduction of magnetic surveys relies on secular variation models.

Spherical harmonic analysis is suitable to model the longest-wavelength part of the geomagnetic field. The IGRF (International Geomagnetic Reference Field) is the accepted global model of the geomagnetic field. It is believed to contain all, or most of, the core field, i.e. the largest part of the field observed at the Earth's surface. The IGRF is given in the form of sets of spherical harmonic coefficients up to and including degree and order 10, except for the most recent main-field models, which extend to degree and order 13, and the predictive secular variation model which extends to degree and order 8. The current 10th generation, revised in 2005 and indicated by the short name IGRF-10, is composed of 12 definitive sets of main field models, ranging from 1945 to 2000.0[4], at 5-year intervals. The latest coefficients are the main field coefficients for 2000.0 and 2005.0 and the secular variation coefficients for 2005.0–2010.0. Since geomagnetic data do not cover all the Earth uniformly, the most suitable method of finding the spherical harmonic coefficients is that based on a least-squares procedure. Because of the irregular geographic distributions of lands and of different economical situations of countries, some regions (e.g. Europe and northern America) are better represented by the IGRF than others (e.g. the oceans).

1.5.3
Geomagnetic Indices

The term magnetic activity is referred to the amplitude variability of magnetic time variations associated with external origin fields. This activity, recorded on the ground by magnetic observatories, is difficult to be exactly quantified. Magnetic indices have therefore been introduced to provide a quantification of the Earth's magnetic field activity level. A few indices are now commonly in use to characterize magnetic activity

---

[4] 2000.0 is a standard use in geomagnetism when referring to maps. It means that all values used to draw the map (which individually have been measured at different times) have been corrected to the same time, i.e. hour 0:00 of January 1st, 2000 (2000.5 means: hour 0:00 of July 1st, 2000).

on a global scale and also on more specific scales. We will describe in what follows the most widely used.

1.5.3.1
K-, Kp- and ap-, Ap-Indices

The K-index summarizes geomagnetic activity generated by solar particle radiation injection into the magnetosphere as recorded at magnetic observatories. The index is expressed by assigning a code, an integer in the range 0 to 9, to each 3-hour Universal Time (UT) interval in a day. Therefore each day is characterized by 8 K-indices. The index for each 3-hour UT interval is determined from the ranges in H and D (scaled in nT), after the removal of the regular expected Sq diurnal variation, generated by solar electromagnetic radiation. In order to remove to an average Sq variation pattern, the five quietest days of the month are used to compute a reference standard variation.

The conversion from a range to a numeric value index, is made using a quasi-logarithmic scale, with the scale values dependent on the observatory's geomagnetic latitude. In fact the same planetary disturbance can show up with different amplitudes at different latitudes. As an example the conversion for L'Aquila, Italy, geomagnetic observatory (geomagnetic latitude 36.3° N) is given in Table 1.2.

In Fig. 1.30 the time variation of the horizontal component H (upper panel) and of declination D (lower panel) of the Earth's magnetic field as observed at L'Aquila observatory on January 23, 2004 is shown. Both magnetic elements (H and D) are measured in nanotesla (nT), while time is measured in minutes. The corresponding series of K-indices for the day is the following: $K = 4, 3, 3, 4, 3, 6, 6, 4$.

The planetary 3-hour-range index Kp is the mean standardized K-index from 13 geomagnetic observatories selected on purpose located between 44° and 60° northern or southern geomagnetic latitude. In this case the scale 0 to 9 is farther subdivided and expressed in such a way to include the thirds of a unit. As an example the symbol 5– represents 4 and 2/3; 5o is 5 and 5+ is 5 and 1/3. Being K- and Kp-indices based on a logarithmic scale a linearized scale index was also introduced. The 3-hourly ap (equivalent range) index is derived from the Kp-index as shown in Table 1.3.

The daily index Ap is obtained by averaging the eight values of ap for each day.

1.5.3.2
AE-Index

The AE-index is an auroral electrojet index obtained from a number (usually greater than 10) of stations distributed in local time in the latitude region typical of the northern hemisphere auroral zone. For each station the north-south magnetic perturbation H is recorded as a function of universal time. A superposition of these data from all the stations enables a lower bound or maximum negative excursion of the H component to be determined; this excursion is called the AL-index. Similarly, an upper bound or maximum positive excursion in H is determined; this is called the AU-index. The difference between these two indices, AU-AL, is called the AE-index. Notice that nega-

Table 1.2. Conversion from a range to a numerical index value for L'Aquila, Italy (geomagnetic latitude 36.3° N)

Difference in nT	K-index
0 – 4	0
4 – 8	1
8 – 16	2
16 – 30	3
30 – 50	4
50 – 85	5
85 – 140	6
140 – 230	7
230 – 350	8
> 350	9

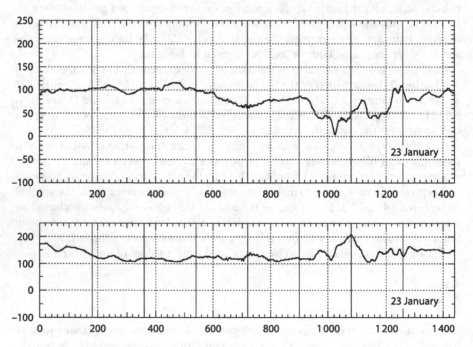

Fig. 1.30. Example of a magnetogram for day January 23, 2004, for L'Aquila Observatory, geomagnetic latitude 36.3° N, and K-indices corresponding values. $K = 4, 3, 3, 4, 3, 6, 6, 4$. H and D are measured in nT and time is measured in minutes

tive H perturbations occur when stations are under a westward-flowing electrical current flowing in the auroral electrojet.

Thus the indices AU and AL give some measure of the individual strengths of the eastward and westward electrojets, while AE provides a measure of the overall

Table 1.3. 3-hourly *ap* (equivalent range) index derived from the *Kp*-index

Kp	ap
0o	0
0+	2
1–	3
1o	4
1+	5
2–	6
2o	7
2+	9
3–	12
3o	15
3+	18
4–	22
4o	27
4+	32
5–	39
5o	48
5+	56
6–	67
6o	80
6+	94
7–	111
7o	132
7+	154
8–	179
8o	207
8+	236
9–	300
9o	400

horizontal current strength. The ordinary time resolution for *AE*-indices is one hour but higher time resolution indices (5 minutes or so) are at times computed for special purposes.

1.5.3.3
Dst-Index

Planetary perturbations characterized by a marked decrease in the *H* (northward) component at mid latitude observatories are called magnetic storms. In order to quantify the *H* depression effect, an hourly *Dst* (Disturbance storm time) index is obtained from magnetometer stations near the equator, but not so close that the E-region equatorial electrojet dominates the magnetic perturbations observed on the ground. At such latitudes the horizontal intensity and the vertical intensity of the magnetic perturbation are dominated by the effects of the magnetospheric ring current. The *Dst*-index is a direct measure of the hourly average of this perturbation. Large negative perturbations in *H* are indicative of an increase in the intensity of the ring current and typically appear on time scales of about an hour. The intensity decrease may take much longer, on the order of several hours or even one or two days. Since during a magnetic storm several isolated or one prolonged substorm signature in the *AE*-index are sometimes observed, specific high time resolution (5 min or so) versions of the *Dst*-index are computed to study the relationship between storms and substorms.

Magnetic indices are in conclusion the quantification of magnetic activity; different indices have been introduced for different purposes. Several applications in geomagnetism make a large use of magnetic indices, however the meaning of geomagnetic indices has changed during the years. For example in the case of *Dst*, it is well known now that this index does not exactly describe the ring current activity only; in fact the decrease in horizontal component at low latitude magnetic observatories is also caused by the existence of other current systems in the magnetosphere and is not only due to the ring current. Nowadays a major objective is the prediction of the state of magnetic activity on Earth and then the prediction of magnetic indices. In space weather the effects begin on the Sun and ultimately affect the Earth, as geomagnetic variations and all related effects. From observations of the Sun and data from interplanetary space-probes the chain of effects is followed and short-term predictions are already available from several space weather services.

Suggested Readings and Sources of Figures

Books

Backus G, Parker R, Constable C (1996) Foundations of geomagnetism. Cambridge University Press, Cambridge, 369 pp
Campbell WH (2001) Earth magnetism. A guided tour through magnetic fields. Harcourt Academic Press, San Diego, CA, USA, 151 pp
Campbell WH (2003) Introduction to geomagnetic fields. Cambridge University Press, Cambridge, 337 pp
Jankowsky J, Sucksdorff C (1996) IAGA guide for magnetic mesurements and observatory practice. Warsaw, 235 pp
Kivelson MG, Russell CT (1996) Introduction to space physics. Cambridge University Press, Cambridge, 568 pp
Lowrie W (1997) Fundamentals of geophysics. Cambridge University Press, Cambridge, 354 pp
Merrill RT, McElhinny MW, McFadden PL (1996) The magnetic field of the Earth: Paleomagnetism, the core and the deep mantle. Academic Press, San Diego, California, 531 pp
Newitt LR, Burton CE, Bitterly J (1996) IAGA guide for magnetic repeat station surveys. Boulder, 112 pp
Tauxe L (1998) Paleomagnetic principles and practice. Kluwer Academic Publishers Group, 312 pp

Articles

Buffett BA (2000) Earth's core and the geodynamo. Science 288:2007–2012
Cain JC, Wang Z, Schmitz DR, Meyer J (1989) The geomagnetic spectrum for 1980 and core crustal separation. Geophys J 97:443–447
De Michelis P, Cafarella L, Meloni A (2000) A global analysis of the 1991 geomagnetic jerk. Geophys J Int 143:545–556
Gonzales WD, Tsuritani B, Clua De Gonzales A (1999) Interplanetary origin of geomagnetic storms. Space Sci Rev 88:529–562
http://solid_earth.ou.edu/ notes/potential/legendre.gif (Copyright 2004, J. Ahern)
Jackson A, Jonkers ART, Walker MR (2000) Four centuries of geomagnetic secular variation from historical records. Philos T Roy Soc A 358:957–990
Kono M, Roberts HR (2002) Recent geodynamo simulations and observations of the geomagnetic field. Rev Geophys 40:4
Lanza R, Meloni A, Tema E (2005) Historical measurements of the Earth's magnetic field compared with remanence directions from lava flows in Italy over the last four centuries. Phys Earth Planet Int 148:97–107
Malin SRC, Barraclough DR (1982) 150th anniversary of Gauss's first absolute magnetic measurement. Nature 297:285
McMillan S and Maus S (2005) International Geomagnetic Reference Field the 10th generation. Earth Planets Space 57(12):1135–1140
Valet JP (2003) Time variations in geomagnetic intensity. Rev Geophys 4:1–44

Chapter 2

Basic Principles of Rock Magnetism

The magnetic properties of matter depend on the fact that elementary particles are provided with their own magnetic moment. In Rutherford's model, the moment of an atom is given by the sum of the magnetic spin moment of the nucleus and of the orbital and spin magnetic moments of the electrons. In many cases, the spin moment of the nucleus can be neglected, because it is smaller by two orders of magnitude than that of the electron. For example, in the case of hydrogen, the ratio between the two moments of the proton and of the electron is 1/660. Considering that electrons occupy orbitals according to precise symmetry rules (for example each orbital can be occupied at most by two electrons rotating in opposite directions), in many cases the magnetic moments of the electrons cancel each other out. The total moment of an atom depends on the number of electrons and on their distribution in the orbitals. In some cases it is zero, in others it is different from zero.

The magnetic moment of a body with finite extension is given by the vector sum of the magnetic moments of its atoms. If the moment of the atoms is $m = 0$, then the total moment of the sample is also nil; if it is $m \neq 0$, then the moment of the sample is still, in the more general case, nil. The vibrations caused by thermal energy cause the orientation of the magnetic moment of an atom to change its direction continuously. The statistical sum of the individual moments is nil, since at any instant they are distributed according to random directions (Fig. 2.1a); at a subsequent instant, the direction of each individual moment has changed, but the set of the distribution is still random and therefore the total moment is still nil (Fig. 2.1b). For a macroscopic sample having a net magnetic moment, a preferential direction of alignment must exist for microscopic moments, which can originate from two different phenomena:

1. an external magnetic field H[1]. In this case, the atomic moments tend to orient themselves in the direction of H (Fig. 2.1c). Why do they *tend*? The alignment of the moment with the external field corresponds to its minimum potential energy, but it is not stable because the thermal activation energy can be enough to disperse the directions of the various moments at random. However, the direction of H is the preferential one and at any instant a certain number of moments is parallel to the field and it gives rise to a total magnetic moment $M = nm\tanh(\mu_0 H / KT)$, n being the to-

---

[1] In Chap. 1 the symbol H was used for the horizontal component of the Earth's magnetic field F, $H = F\cos I$. Both F and H are magnetic inductions and measured in T. From now on, we shall use the symbols H for the strength of a generic magnetic field, H_E for the strength of the Earth's magnetic field, $F = \mu_0 H_E$. Both H and H_E are measured in A m^{-1}.

tal number of atoms, $K = 1.38 \times 10^{-23}$ Joule/°K Boltzman's constant and T absolute temperature. Clearly, then, the magnetic moment of a sample may vary from $M = 0$, in the absence of external magnetic field, to $M = nm$, if the field is so intense as to orient all atomic moments. In the more general case $M \neq 0$, we can define the magnetization vector as the magnetic moment per unit of volume $J = M / V$; as we shall see, this vector is proportional to the external magnetic field, $J = \kappa H$, and the proportionality constant is called magnetic susceptibility.

2. magnetic interaction forces between electrons. Some crystalline substances have a very dense structure and consequently a strong interaction between the magnetic spin moments of the electrons, which causes a spontaneous alignment of the moments themselves and hence a resulting magnetization J even in the absence of an external magnetic field.

Fig. 2.1. Atomic magnetic moments in a solid. Moment orientation is random (**a**) and it varies from one instant to the other (**b**). The net magnetic moment is zero. In the presence of an external magnetic field H (**c**), a fraction of the moments is aligned with the field and the solid acquires an induced magnetization J_i

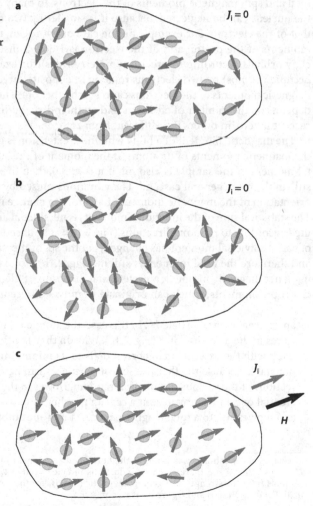

2.1
Magnetic Properties of Solids

The magnetic field in a generic point is described by the equation

$$B = \mu_0 H + \mu_0 J \tag{2.1}$$

where B is magnetic induction (expressed in T), H magnetic field strength (A m^{-1}), J magnetization (A m^{-1}) or magnetic moment per unit of volume. In vacuum, J is nil, in matter its properties depend on those of the elementary particles, according to which substances are subdivided in three categories: dia-, para- and ferromagnetic.

2.1.1
Diamagnetism

Atoms' electrons can be considered as particles of mass m_e and electrical charge $-e$, which move along a circular orbit with angular velocity ω and therefore have an orbital magnetic moment m (Fig. 2.2a). In the presence of an external magnetic field B, electrons are subjected to the Lorentz force $F_L = e\nu \times B$, which, depending on the direction of rotation, is subtracted from or added to the centripetal force of electrostatic origin F_e, causing the angular velocity decreases or increases (Fig. 2.2b) of the quantity $\Delta\omega = eB/2m_e$ (Larmor precession). The corresponding change in the orbital magnetic moment Δm is the same in both cases, since the moment decreases when its direction is the same as that of B and increases when it is opposite. Any substance subject to an external magnetic field acquires a magnetic moment opposite to the field direction.

If all the orbitals in a molecule are complete, the magnetic moments of the electrons cancel each other out and the total moment is nil. These substances are called diamagnetic and in the presence of an external magnetic field, due to Larmor's precession, acquire a weak induced magnetization $J_i = \kappa H$ in the opposite direction to that of the field, where κ is magnetic susceptibility, a negative numeric (i.e. dimensionless) constant (Fig. 2.3) typical of each substance and independent of temperature

$$\kappa = -(NZ\mu_0 e^2 \langle r^2 \rangle)/6m_e \tag{2.2}$$

where N is the number of atoms per unit of volume, Z the number of electrons of the atom, $\langle r^2 \rangle$ the mean square radius of the electron orbit. In the case of minerals, the value of κ is in the order of -10 µSI (1 µSI $= 10^{-6}$ SI unit). Larmor's precession is a fundamental property of matter, but the magnetization J_i it produces is so weak as to be macroscopically observable only when it is the sole effect of the magnetic field, as is the case in diamagnetic substances, and to be obscured by other effects in para- and ferromagnetic substances.

2.1.2
Paramagnetism

Paramagnetic substances are those whose molecules have their own magnetic moment m: not all orbitals are complete, and unpaired electrons have a free magnetic spin

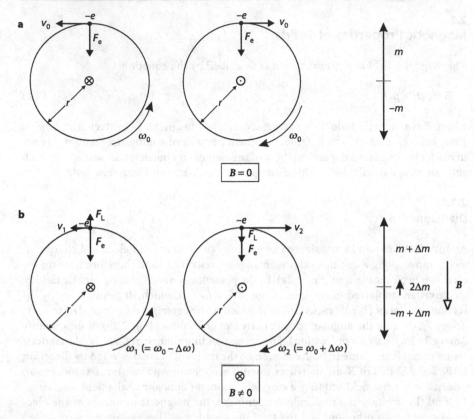

Fig. 2.2. The two electrons of an orbital circulate in opposite directions; **a** in the absence of a magnetic field, the magnetic moments (the left one goes into the page, the one on the right comes out of it) are equal and opposite and the net effect is cancellation; **b** the presence of a field B (going into the page) produces a Lorentz force F_L which increases or decreases angular velocity ω_0, according to the direction of rotation: one moment increases, the other one decreases and their resultant $2\Delta m$ is opposite to the field

moment. In the case of crystalline solids, the thermal energy of the atoms causes continuous lattice vibrations, and the orientation of the moments varies randomly from one instant to the other (Fig. 2.1). The resultant is therefore nil. In the presence of an external magnetic field, the dispersing effects of temperature are contrasted by the torque exerted by the field on individual magnetic moments, and an equilibrium is reached in which the degree of alignment depends on the ratio between magnetic and thermal energy, $\alpha = mB / KT$, to which corresponds an induced magnetization

$$J_i = NmL(\alpha) \tag{2.3}$$

where $L(\alpha) = \coth(\alpha) - 1/\alpha$ is Langevin function (Fig. 2.4). Equation 2.3 shows that magnetization is positive, hence in the same direction as the external field, and de-

Fig. 2.3. Induced magnetization J_i versus magnetic field H in dia- and paramagnetic substances

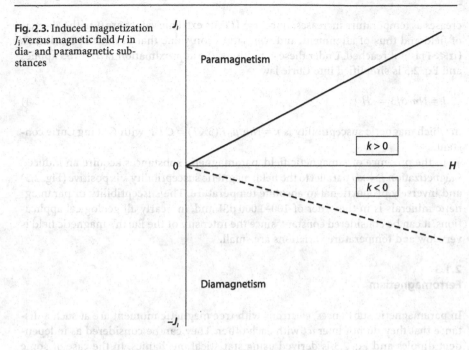

Fig. 2.4. The Langevin function. $L(\alpha) \approx \alpha/3$ for $\alpha = mB/KT \ll 1$

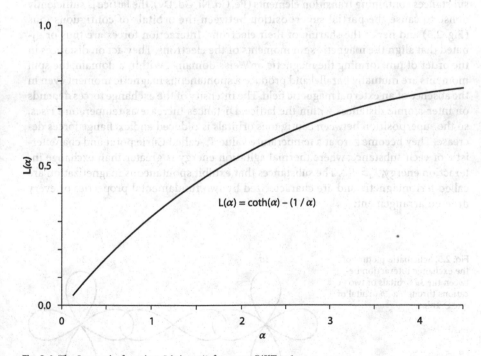

creases as temperature increases, since $\alpha \propto 1/T$. Its extreme values are 0, in the absence of field and thus of alignment, and Nm, saturation value that in normal conditions ($\alpha \ll 1$) is not reached. Under these conditions, the approximation $L(\alpha) \approx \alpha/3$ applies and Eq. 2.3 is simplified into Curie law

$$J_i = Nm\,\alpha/3 = \kappa H \tag{2.4}$$

in which magnetic susceptibility is $\kappa = Nm^2\mu_0 / (3KT) = C / T$, with C being Curie constant.

In the presence of a magnetic field, paramagnetic substances acquire an induced magnetization $J_i = \kappa H$ parallel to the field; magnetic susceptibility κ is positive (Fig. 2.3) and inversely proportional to absolute temperature. The susceptibility of paramagnetic minerals is in the order of 100–1 000 µSI and, in nearly all geological applications, it can be considered constant, since the intensity of the Earth's magnetic field is very low and temperature variations are small.

2.1.3
Ferromagnetism

In paramagnetic substances, electrons with free magnetic moment are at such a distance that they do not interact with each other. They can be considered as independent dipoles and Eq. 2.3 is derived using statistical mechanics. In the case of some substances containing transition elements (Fe, Co, Ni, Gd, Dy), the lattice is sufficiently dense to cause the partial superposition between the orbitals of contiguous ions (Fig. 2.5) and hence the sharing of their electrons. Interaction forces are thus originated that align the magnetic spin moments of the electrons. They act on distances in the order of µm, forming the magnetic or Weiss domains. Within a domain, the spin moments are mutually parallel and produce a spontaneous magnetic moment even in the absence of an external magnetic field. The intensity of the exchange forces depends on inter-atomic distances within the lattice. Distances increase as temperature rises, so the superposition between contiguous orbitals is reduced and exchange forces decrease. They become zero at a temperature value T_c, called Curie point and characteristic of each substance, where thermal agitation energy is greater than exchange interaction energy, $E_t > E_{ex}$. The substances that exhibit spontaneous magnetization are called ferromagnetic and are characterized by two fundamental properties of every ordered arrangement:

Fig. 2.5. Schematic picture of the exchange interaction between the $3d$ orbitals of two Fe cations through a $2p$ orbital of one O anion

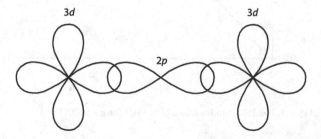

- A change in the disposition of a magnetic domain requires energy, in the absence of which spontaneous magnetization tends to be maintained over time. For this reason, the magnetization of ferromagnetic substances is also called permanent.
- The iso-orientation of the spins is linked to the characteristics and symmetry of the lattice and it spontaneously occurs along some directions called *easy*. Ferromagnetic substances are thus intrinsically anisotropic.

Depending on the substance, exchange forces can act directly between two contiguous cations or through an anion interposed between them. In the first case, we have ferromagnetic substances *sensu stricto (s.s.)*, whose spin moments are all mutually parallel and concordant and impart a total magnetic moment to the domain (Fig. 2.6a). In the second case, the spin moments of contiguous cations are antiparallel and thus form two sub-lattices, magnetized in opposite directions. Depending on the number of cations involved and on the crystalline structure, three different cases are possible. In antiferromagnetic substances, the two sub-lattices have the same intensity and hence their magnetizations, with equal modulus and opposite directions (Fig. 2.6b), cancel each other out. The overall domain has zero magnetic moment. In ferrimagnetic substances, the two sub-lattices have different magnetizations (Fig. 2.6c): their resultant is no longer nil and the domain has a magnetic moment. Lastly, imperfect (also known as canted or parasitic) antiferromagnetic substances have their two sub-lattices not exactly antiparallel, due to the presence of impurities, lattice flaws and vacancies, or the phenomenon called spin canting (Fig. 2.6d). The resultant of the two magnetizations is slightly different from zero and the domain has a weak magnetic moment. For all these substances, ferromagnetic *sensu lato (s.l.)*, the temperature beyond which the ordered state cannot survive is generically known as Curie temperature T_c, but when referring specifically to antiferromagnetic substances the correct term is Néel temperature T_n.

From the above, it is apparent that chemical composition and crystalline structure play a fundamental role in determining the properties of ferromagnetic substances and the enormous difference between artificially produced substances and rocks is clear. In the first case, the substance has well defined magnetic properties and manufacture processes try to attain the highest possible degree of homogeneity, which almost never exists in the case of natural minerals. Small chemical variations within isomorphic series, different oxidation state, impurities and lattice flaws, coupled with variations in grain shape and size all together concur to cause the magnetic properties of minerals and rocks to be extremely complex and variable.

Returning now to ferromagnetic substances in general, let us consider a polycrystalline sample, which is subdivided into many magnetic domains. Every domain will have its own spontaneous moment (Fig. 2.7) and transition to contiguous domains is not discontinuous, but takes place gradually through a domain or Bloch wall, i.e. a transition region with a thickness of 10^{-1}–10^{-2} μm in which spins progressively rotate from the direction of one domain to that of the adjacent domain (Fig. 2.8). The different orientations of the moments of the various domains cause their resultant, i.e. the macroscopically observable moment, to have a random value, which may even be nil. This is the reason why a piece of ferromagnetic substance like iron can lack macroscopic magnetization. Subjecting the sample to an external magnetic field H, the walls of the domains move to favor the growth of the domains with magnetization in the

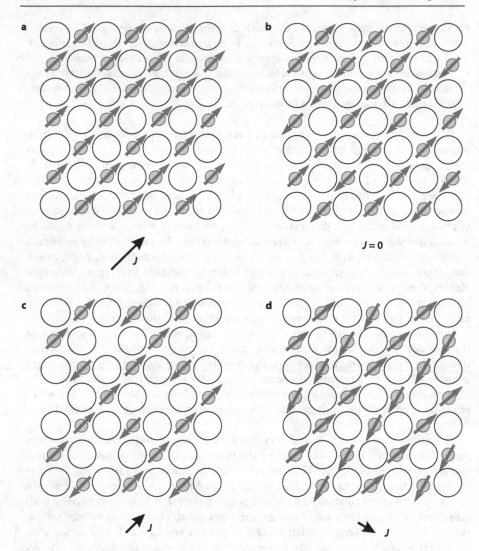

Fig. 2.6. Schematic picture of spin moment alignment in substances; **a** ferromagnetic *s.s.*; **b** antiferro-magnetic; **c** ferrimagnetic; **d** canted antiferromagnetic (modified after O'Reilly 1984)

direction of the field. If the value of H is small, movements are reversible and on removal of the field the domains return to their previous equilibrium situation (Fig. 2.9). The external field then produces an induced magnetization that is similar to the one observable in paramagnetic substances and in this case, too, it is proportional to field intensity, provided it is small, $J_i = \kappa H$. Magnetic susceptibility κ is positive and in the case of ferromagnetic minerals it is in the order of 10^4–10^7 μSI. For high values of H, the growth of the domains parallel to the field continues passing through a discontinuous series of equilibrium situations (Fig. 2.9), reached with irreversible jumps

Fig. 2.7. Domain arrangement in a polycrystalline ferromagnetic material formed in the absence of a magnetic field. Each domain is spontaneously magnetized in its own easy direction

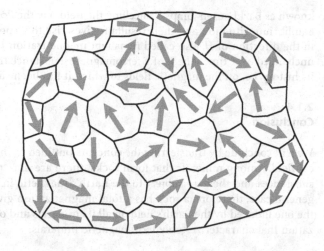

Fig. 2.8. Progressive spin moment rotation through a domain wall

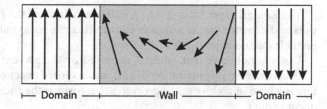

├── Domain ──┤ ├──────── Wall ────────┤ ├── Domain ──┤

Fig. 2.9. Energy of a domain wall as a function of its position. The wall that separates two domains is in a potential energy minimum (A). An external field causes the left side domain to grow. If the new position of the wall (B) lies between two potential maxima (1-2), then the displacement is reversible and the wall returns to (A) when the field is removed. If the wall crosses a maximum (2), on removal of the field the wall spontaneously migrates to the new minimum (C) (modified after Stacey and Banerjee 1974)

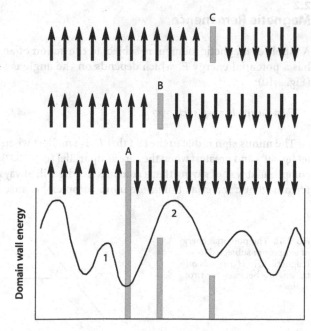

known as Barkhausen jumps. Cancelling the field out, the domains remain in the new equilibrium situation and their resultant gives rise to a spontaneous magnetization in the direction of H, also called remanent magnetization J_r, which tends to remain unchanged over time. The J_r of a ferromagnetic substance therefore also depends on its history, i.e. on the magnetic fields to which it has been subjected.

2.1.4
Conclusion

We will conclude by transferring the concepts outlined in this section to the geological field. Among minerals that form rocks, there are dia-, para- and ferromagnetic substances and they are subject to the Earth's magnetic field, $F = \mu_0 H_E$. In the most general case, therefore, a rock has a total magnetization given by the vector sum of the one induced by the Earth's field on all its minerals and of the remanent magnetization that characterizes only ferromagnetic minerals:

$$J_{tot} = J_i + J_r = \kappa H_E + J_r \tag{2.5}$$

The Königsberger ratio $Q = J_r / J_i$ indicates which of the two magnetizations prevails and depends on the minerals constituting the rock and their content as well as on their magnetic properties.

Lastly, we observe that, whilst J_i is caused by the present-day field, J_r, also called natural remanent magnetization (NRM), was acquired over geological times and therefore is a kind of archive of the processes that formed the rocky body and of its subsequent history.

2.2
Magnetic Remanence

A dipole of magnetic moment m subject to the action of an external magnetic field H has a potential energy E_H which depends on the angle φ formed by the two vectors (Fig. 2.10):

$$E_H = -\mu_0 m \cdot H = -\mu_0 m H \cos\varphi \tag{2.6}$$

The minus sign is due to the fact that E_H is smallest when the two vectors are parallel ($\varphi = 0°$) and greatest when they are antiparallel ($\varphi = 180°$). Equation 2.6 is a bit more complicated way of saying that a compass needle will always point north. In a grain of ferromagnetic material, things are more complex because there are many magnetic

Fig. 2.10. The potential energy of a dipole m subjected to a magnetic field H depends on the angle φ between the two vectors

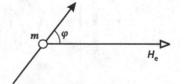

moments, interferring with each other; moreover, they are not a purely conceptual pair of point-like magnetic poles, but they originate from complex atomic phenomena. In Sect. 2.1 we considered a highly simplified model, valid as a first approximation for paramagnetic substances. There were two energies at play: one linked to the magnetic field and one to temperature. In the case of ferromagnetic substances, the energy balance is more complicated.

2.2.1
Magnetic Energies

Let us consider a uniformly magnetized grain (J = constant), which generates, in the surrounding space, a magnetic field schematically indicated with the field lines in Fig. 2.11. The elementary dipoles are all mutually parallel and the north end of a dipole faces the south end of the contiguous dipole. Their effects cancel each other out, but the ends situated at the surface of the grain are clearly not compensated and they are the ones that originate the magnetic field in the surrounding space. The situation can be described in similar terms to those used in electrostatics: the field produced by the grain is caused by a distribution of surface magnetic charges; those corresponding to north are positive, those corresponding to south are negative. The interior of the grain does not contribute and can therefore be ignored. But field lines irradiate from each positive charge in all directions and converge from all directions towards each negative charge. Therefore, they also must pass through the internal part of the grain, where they are antiparallel to J. Hence, there is also an internal field, which is called demagnetizing field H_d because it is opposite to J and its effect is to reduce the overall magnetization of the grain. The demagnetizing field is given by the relationship

$$H_d = -NJ \tag{2.7}$$

where the demagnetizing factor N is a scalar that depends on both the shape of the grain and its direction relative to J. The existence of H_d entails an internal magneto-static energy E_m, which in the case of uniform magnetization is given by a relationship similar to (Eq. 2.6), where V is the volume of the grain and the factor 1/2 takes account of the fact that every north pole is associated to a south pole

$$E_m = -1/2 \, \mu_0 VJ \cdot H_d = 1/2 \, \mu_0 VNJ^2 \tag{2.8}$$

Considering an ellipsoidal grain, the value of N in a certain direction can be interpreted as the percentage of surface on which magnetic charges are located, and calling $a > b > c$ the three axes of the ellipsoid, we will have the relationship $N_a + N_b + N_c = 1$. In the case of a sphere, for evident symmetry reasons $N_a = N_b = N_c = 1/3$ (Fig. 2.12), while in the cases of a revolution ellipsoid, $N_a < N_b = N_c$ if prolate, $N_a = N_b < N_c$ if oblate. In the limit case of a very thin needle, $N_a \approx 0, N_b = N_c = 1/2$, in the case of a very thin disk $N_a = N_b \approx 0, N_c \approx 1$. The experimental observation that it is easier to magnetize an iron bar parallel to its length, more difficult transversely, is explained by the fact that the internal magnetostatic energy E_m in the first case is smaller than in the second.

So far we have considered the grain as homogeneous, but in fact its structure is crystalline and its physical properties are anisotropic. Ferromagnetism is caused by

Fig. 2.11. Demagnetizing field in a uniformly magnetized grain. All elementary dipoles are parallel (**a**) and the N and S poles of adjacent dipoles compensate each other; only the poles along the outer surface are not compensated. The magnetic field produced by the grain (**b**) is equivalent to that of a set of point-like poles: the field lines irradiate from the N poles and converge into the S poles, but they do so in all directions and hence also inside the grain (**c**), giving rise to an internal field, which is demagnetizing because it is opposite to the polarization that causes it (from O'Reilly 1984)

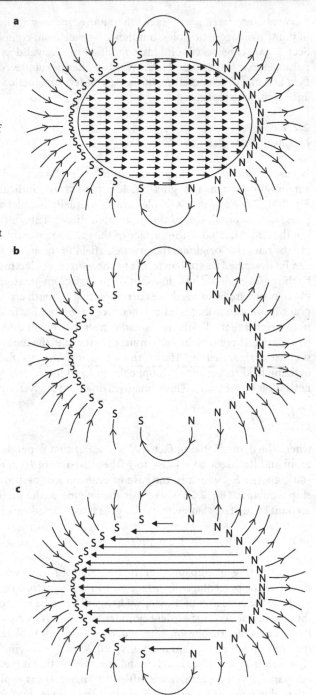

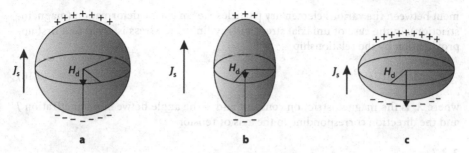

Fig. 2.12. Demagnetizing field H_d and saturation magnetization J_s of; **a** a sphere; **b** a prolate ellipsoid; and **c** an oblate ellipsoid

Fig. 2.13. The coupling between the orbital magnetic moment and the spin moment of an electron can be schematized considering the electron as the origin of a frame of reference. The nucleus orbits around the electron and generates a magnetic field H which interacts with m (from O'Reilly 1984)

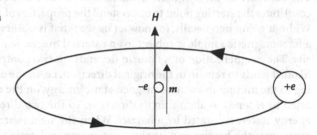

interaction forces between the spins of the electrons of certain orbitals, but the electron also has an orbital magnetic moment and two magnetic moments can only interact with each other. The orbital-spin coupling (Fig. 2.13) does not generate ferromagnetism, but is able to influence it. The disposition of the orbitals depends on the characteristics of the lattice and can give rise to directions along which the electron spins tend to orient preferentially, called easy magnetization directions. These directions therefore correspond to minimum values of magneto-crystalline anisotropy energy, E_a. In the case of crystals with uniaxial symmetry, the following relationship applies

$$E_a = K_u V \sin^2 \theta \tag{2.9}$$

where K_u is a typical constant of each substance and θ the angle the magnetization J forms with the easy direction of magnetization.

The anisotropy of a crystal depends on the ordered disposition of the nuclei and electrons and the various forces exerted between these elementary particles depend first of all on their mutual distances. A magnetized grain subject to an external stress σ, for instance of tectonic origin, undergoes a deformation that varies the inter-atomic distances, hence the symmetry, hence magneto-crystalline anisotropy. Magnetization is modified (piezomagnetism) and part of the elastic energy due to stress is transformed into magnetic energy. Similarly, if a grain is magnetized under the action of an external magnetic field, the spin-orbit coupling causes a small rotation of the orbital, which entails a variation in the electrostatic forces and hence a relative movement between the various elementary particles, i.e. an elastic deformation (magneto-striction). In the case of uniaxial stress, energy linked to stress is given as a first approximation by the relationship

$$E_\sigma = 3/2\, \lambda_s \sigma V \sin^2\theta \qquad\qquad (2.10)$$

where λ_s is the magnetostriction constant and θ the angle between magnetization J and the direction corresponding to the axis of tension.

2.2.2
Magnetic Hysteresis

The existence of the magnetic domains and of the associated magnetostatic energy constitutes the starting point to understand the properties of ferromagnetic substances. Without going into details, for now let us see what is observed macroscopically when a ferromagnetic sample is subject to an external magnetic field H with variable intensity. The magnetization of a domain depends on two contrasting effects. On the one hand, it tends to remain in the original direction, i.e. in the easy direction corresponding to the minimum value of magnetostatic energy, on the other hand the torque exerted by H tends to align it in its direction. To the two directions correspond to two energy states, separated by a barrier. When the field is strong enough, the domain overcomes the barrier and its spontaneous magnetization is recreated in the direction of the field.

Let us suppose a field $H = 0$ as the initial state: each of the domains of the sample is oriented according to its own easy direction and all together their resultant is $J = 0$. With a small intensity of H, the displacement of the walls favors the growth of the domains magnetized in directions close to that of the field and the sample thus acquires a reversible magnetization. Further increasing H, some domains start rotating their direction of magnetization (Fig. 2.14a) with a discrete transition from the state of lowest magnetostatic energy to a state with greater energy, but favored by the fact that the rotation reduces energy linked to the influence of H. Magnetization intensity J increases, with a marked growth at the beginning which then tends to taper off (Fig. 2.14b). As the domains become progressively aligned, the number of those whose direction is different from H is reduced, until for a certain value H_s, called saturation field, all domains are parallel and their resultant reaches saturation magnetization J_s, i.e. the greatest possible value (Fig. 2.14a,b). An additional increase in the field no longer has any effect, since there are no domains left to orient. Letting H decrease now, the interaction energy between field and domains decreases: some domains return to the easy magnetization situation and J gradually decreases (Fig. 2.14b). When the applied field reaches nil, $H = 0$, not all domains are able to overcome the energy barrier separating them from the easy magnetization state and a certain number remains oriented in the direction H had. The sample has a residual magnetization $J_{rs} \neq 0$, called saturation remanence. To reduce J_{rs} to zero, H must be made to grow in the opposite direction, so that the domains start to become magnetized in the new direction and the corresponding magnetization $-J$ is subtracted from the previous one. The coercive force $-H_c$ is the field value at which $J = 0$ is measured. The coercivity of saturation remanence is the value $-H_{crs}$ necessary to reduce J_{rs} to zero,

Fig. 2.14. Magnetic hysteresis;
a the domains whose magnetization is concordant with an
increasing field H first grow at
the expense of the others, then
rotate their magnetization in
the direction of H (*Segment 1
of the hysteresis curve*);
b hysteresis loop. Symbols: H_s,
J_s = saturation field and magnetization; J_{rs} = saturation
remanence; H_c = coercive force;
H_{crs} = coercivity of saturation
remanence. H_{crs} is the field that
must be applied so that, after
the removal of the field itself
(*Segment 2 of the curve*), $J = 0$.
If saturation is not reached, the
loop is smaller and the remanence and coercivity values are
lower, $J_r < J_{rs}$ and $H_{cr} < H_{crs}$

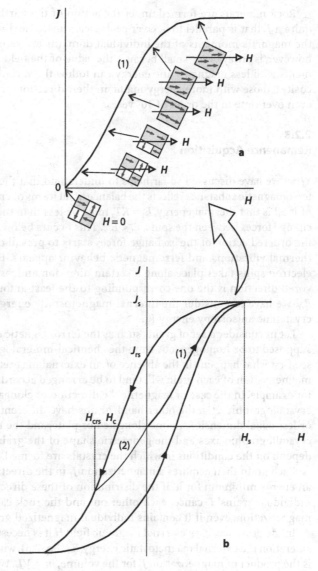

i.e. to measure a nil remanence ($J_r = 0$) after removing the applied field. Continuing
to increase the modulus of $-H$, $-J$ also increases until reaching the values $-H_s$ and $-J_s$,
which have the same meaning as before. Returning in a similar way to the value $+H_s$
one obtains a curve that is symmetrical to the previous one and the hysteresis loop is
closed.

If the experiment is carried out reaching a maximum field value less than saturation, $H < H_s$, the hysteresis curve continues to be a loop; it is just narrower, because
saturation is not reached. In this case, remanent magnetization is $J_r < J_{rs}$ and reducing
it to zero requires a coercivity of remanence $H_{cr} < H_{crs}$.

Rock minerals are formed under the action of the Earth's field H_E and hence acquire a J_r that is parallel to it. Over geological times, the direction of H_E changes and the magnetic moments of the individual domains are subjected to a torque which, however, is always very small because the value of the field strength H_E is small. Moments with less magnetostatic energy can follow the variations in H_E (magnetic viscosity), those with more energy maintain their direction of magnetization unaltered even over time in the order of 10^9 years.

2.2.3
Remanence Acquisition

What we have discussed so far helps to understand that the magnetization state of a ferromagnetic substance reflects the balance of all forms of energies involved. The basis of it all is that thermal energy, $E_t = KT$, must be less than the energy linked to the exchange forces between the spins, $E_t < E_{ex}$. This occurs below the Curie point T_c, where the ordered action of the exchange forces starts to prevail on the disorder caused by thermal vibrations and ferromagnetic behavior appears. But the coupling between electron spins takes place along a certain direction and, as always in nature, the favored direction is the one corresponding to the least amount of energy. Neglecting E_σ, we have to consider the internal magnetostatic energy, E_m, and the magneto-crystalline anisotropy energy E_a.

Let us consider a set of grains such as the ferromagnetic crystals present in a rock, supposed to be homogeneous from the chemical-mineralogical viewpoint, and let us see first what happens in the absence of an external magnetic field. The spontaneous magnetization of each grain will tend to be arranged according to the easy directions: for example, in the case of magnetite the direction of elongation of the grain or [111] crystallographics. On the other hand, grains have different shapes, some with more or less equal dimensions, some elongated or plate, and the relationships between the crystallographic axes and the geometric shape of the grain may vary, because they depend on the conditions in which the crystals are formed.

Each grain then acquires a magnetization J_r in the direction which corresponds to an energy minimum for it. If the distribution of these directions in space is random, individual grains' J_r cancel each other out and the rock can have zero macroscopic magnetization, even if it contains individual magnetized grains.

In the presence of an external magnetic field H it is necessary also to take into consideration the external magnetostatic energy E_H (Eq. 2.6), where the magnetic moment is the product of magnetization J_r for the volume, $m = VJ_r$. Whilst the needle of a compass rotates to orient itself parallel to the field and to minimize its potential energy, the rotation of J_r is hindered by the demagnetizing field H_d and by magneto-crystalline anisotropy, which tend to maintain J_r in the easy direction. The grain is magnetized parallel to the direction of the field only if E_H is greater than the energy barrier, for example magnetostatic energy E_m, which separates the direction of the field and the easy one

$$E_H > \Delta E_m = 1/2\ \mu_0 V(N_H - N_e)J_r^2 \qquad (2.11)$$

where N_H and N_e are the demagnetizing factors in the direction of the external field and in the easy one, respectively. If we consider a set of grains as the one in the previous case, the fraction of grains for which the condition (Eq. 2.11) is met has a coherent remanence J_r, directed in the sense of H. These remanences sum up and they give a macroscopic remanent magnetization to the rock (Fig. 2.15). Magnetization intensity is the greater, the greater the external magnetic field.

To summarize: when the ferromagnetic crystals of a rock are formed, they acquire a spontaneous magnetization which is directed according to the easy directions close to the direction of the external field, i.e. the Earth's field H_E. Given that the strength H_E is low, the fraction of magnetized grains concordant with H_E is small, but sufficient to impart to the rock a natural remanent magnetization (NRM) parallel to the Earth's field. The NRM is maintained over time, unless some phenomenon provides the rock with sufficient thermal or external magnetostatic energy, E_t or E_H, to overcome the energy barriers internal to the grains and to produce a new magnetization state.

2.2.4
Magnetic Domains

In the previous section we discussed ferromagnetic grains without saying anything about their dimensions. However, these have an essential role and they lead us to consider the magnetic domains, which have already been briefly mentioned. The domains are the link between the world of atoms, where elementary magnetic phenomena occur, and the macroscopic world of crystals and they can be seen as the constitutive elements that determine the magnetic properties of a rock.

Let us consider a ferromagnetic grain in which the spin moments of the electrons are all parallel to each other: the results will be a magnetization J_r in the easy direction and an internal magnetostatic energy E_m, proportional to the volume of the grain (Eq. 2.8). As volume grows, E_m increases and it continues to increase until at a certain

Fig. 2.15. The ferromagnetic grains of a rock acquire their spontaneous magnetization along the easy directions close to Earth's field F. Their resultant J_r is the natural remanent magnetization (NRM)

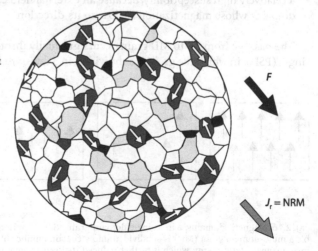

F

$J_r = $ NRM

point in the total energy balance of the grain, it can be convenient to reduce E_m: the grain subdivides into two or more parts in which the alignment of the spin moments is antiparallel (Fig. 2.16) and hence the total magnetostatic energy is reduced. Each part with coherent spin alignment is a domain: in the first case, the grain is single domain (SD), in the second one multiple domain (MD). We have already stated that the transition between contiguous domains occurs with a gradual rotation of the spins, which therefore are no longer parallel to the easy direction: there is an energy associated with the walls of the domain.

The differences between the two types of grains are considerable and the presence of SD or MD grains gives very different magnetic properties to a rock. A SD grain is characterized by

- a strong magnetization, in principle equal to the saturation magnetization J_s, because all its spin moments are parallel;
- a high coercivity, because an external magnetic field must cause all the spontaneous magnetization to rotate away from the easy direction. For example, the hysteresis loop relating to a field H parallel with the easy direction has a rectangular shape (Fig. 2.17) because only the two directions, parallel and antiparallel to the field are possible, with a discontinuous transition for $H = H_c$, where $H_c = 2\,K_u\,/\,J_s$ is microscopic coercivity;
- a relatively low susceptibility, in theory equal to zero in the easy direction because since all the spins are parallel the application of an external field entails no change.

On the other hand, a MD grain has

- a magnetization $J < J_s$, since magnetization is coherent in each domain, but its direction is variable between the domains.
- a smaller coercivity, since an external field gradually aligns the various domains in its direction. The hysteresis curve is similar to the one seen previously (Fig. 2.14).
- a relatively high susceptibility, because an external field can favor the growth of the domains whose magnetization is close to its direction.

The passage from SD to MD grain occurs gradually through a state, called pseudo-single (PSD), in which the grain is subdivided in a few domains. In this case, the inter-

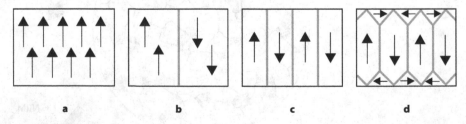

     a   **b**   **c**   **d**

Fig. 2.16. Magnetic domains; **a** in a single domain grain (SD) all spin moments are mutually parallel; **b,c** a multi-domain grain (MD) is subdivided into a certain number of magnetized domains to minimize magnetostatic energy, which is further reduced if closure domains (**d**) are formed along the surface of the grain

Fig. 2.17. Hysteresis loop of an SD grain magnetized parallel to the applied field: a discontinuous transition between the two values $\pm J_s$ occurs in correspondence with the values $\pm H_c$

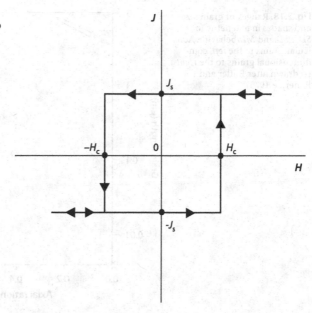

action between the walls and the surface of the grain limit the motion capability of the walls and the magnetic characteristics are similar to those of a SD grain. The magnetic behavior of a grain depends on various factors: type of mineral, dimensions, shape. In the case of magnetite (Fig. 2.18) the MD behavior occurs for dimensions >1–10 μm, the PSD between 0.1 and 3–5 μm, the SD for dimensions <1–0.03 μm. The smaller values refer to equi-dimensional grains and the larger ones to acicular grains, whose elongated shape hampers the birth and the movement of walls. Ultra-fine grains have a particular behavior, called super-paramagnetic (SP). As dimensions decrease, the various magnetic energies considered in the previous section decrease because they are proportional to the volume. Consequently, the energy barrier separating the two possible magnetization states of a SD grain also decreases. Below a critical threshold (≈0.03 μm for magnetite), the energy barrier is comparable to the thermal activation energy KT even at ambient temperature, and the probability that thermal energy is enough to overtake the barrier and reverse the direction of magnetization becomes very high. The grain continues to exhibit the typical alignment of ferromagnetism, but its direction changes continually even over a few seconds. The SP and paramagentic behaviors are similar: an external field orients the magnetization of the grain in its direction, but once the field is removed the magnetization decays in a short time.

2.2.5
Remanence vs. Time

The distinctive characteristic of ferromagnetic substances is that they exhibit a permanent magnetization. So far, we have dealt with the various forms of magnetic energy and discussed their influence in determining the coherent orientation of the spin

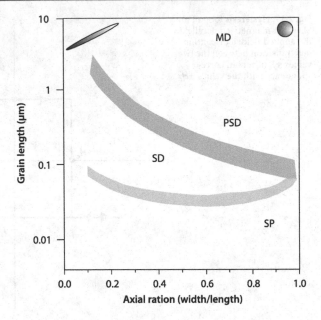

Fig. 2.18. Ranges of grain sizes and shapes in magnetite for *SP*, *SD*, *PSD* and *MD* behavior. Acicular grains to the *left*, equidimensional grains to the *right* (redrawn after Butler and Banerjee 1975)

moments, which constitutes the essence of ferromagnetism. However, we have neglected the adjective *permanent*, i.e. the time aspect of the phenomenon, which essentially depends on the thermal activation energy KT. The latter is subject to continuous alteration over time: therefore, there is a certain probability that the thermal energy of a grain will overcome the energy barrier which keeps the magnetization locked and that the magnetization will change spontaneously, in similar fashion to what has been mentioned above for super-paramagnetism. Since the process is probabilistic, it can be examined statistically. The simplest case is that of a set of SD grains, all having the same shape, dimensions, and mineralogical characteristics. Let us suppose that these grains are subject to a magnetic field that magnetizes all of them in the same direction: the total magnetization of the set is J_0. Once the field is removed, the magnetization decays over time according to the law proposed by Néel

$$J(t) = J_0 \exp(-t/\tau) \tag{2.12}$$

where τ is the relaxation time, i.e. the time that must elapse for the magnetization to be reduced by a factor $1/e$: $J(\tau) = J_0 / e$. The relaxation time depends on the relationship between the energy barrier that opposes the rotation of J and the thermal energy

$$\tau = 1/C \exp(\mu_0 V H_c J_s / 2\, KT) \tag{2.13}$$

where C is a frequency in the order of $10^9\ \mathrm{s}^{-1}$ linked to the time on which the thermal fluctuations occur, V is the volume of the grain, H_c its coercive force and J_s its saturation magnetization. The various parameters that influence τ all appear under exponential: a small change in them entails a large change in τ. Considering temperature, immediately below the Curie point thermal energy can be very high and the energy barriers of the grains low: τ is a few seconds and the grains are super-paramagnetic. When T decreases, thermal energy decreases, energy barriers increase and magnetization is preserved over longer times. The blocking temperature T_b is the temperature below which the probability that thermal energy overcomes the locking energy barrier is less than 1. Magnetization is stable, i.e. it undergoes no substantial changes with any further decreases in temperature and it can be changed only by a more or less intense magnetic field. The value of T_b is associated to a value of τ: laboratory measurements allow directly to check magnetization stability for a certain value T_b on times in the order of $\tau = 100$–$1\,000$ s. The result can be extrapolated to a geological situation considering the T_b corresponding to time intervals of the order of 10^6–10^9 years, i.e. $\tau \approx 3 \times 10^{13}$–$10^{16}$ s, which will obviously be lower. The lower the temperature, in fact, the lower the probability that the thermal activation energy of a domain is able to overcome its energy barriers, and a lower probability means that a longer time is required for the event to occur. Figure 2.19 represents the relationship between T_b and τ for SD magnetite.

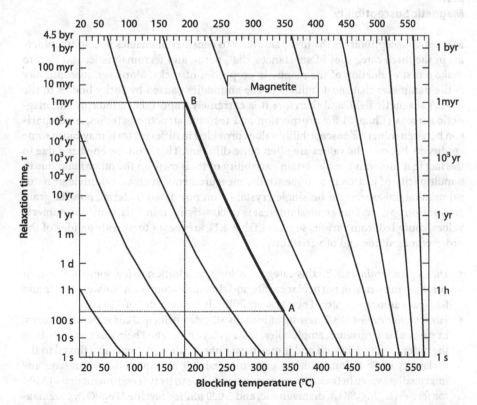

Fig. 2.19. Relaxation time and blocking temperature in SD magnetite. The relaxation time increases with decreasing temperature, because the probability that a domain may change its own spontaneous magnetization decreases. A set of grains having $\tau = 5$ minutes for $T = 340\,°C$ (*A*) shifts to $\tau = 1$ Myr for $T = 180\,°C$ (*B*) (modified after Pullaiah et al. 1975)

A set of SD grains therefore maintains its magnetization for a period that depends on their characteristic τ. The fact that this period can greatly exceed 10^9 years indicates that the rocks are able to maintain their remanent magnetization over times of the same order of magnitude as geological times. The theoretical model for PSD and MD grains is more complex and their relaxation time, especially in the MD case, is shorter. The characteristic values of T_b of a rock can, however, be evaluated experimentally over laboratory times and, with all due caution, extrapolated over longer times.

2.3
Magnetic Properties of Minerals

The magnetic properties of minerals are deeply linked to their mineralogical characteristics and nearly imperceptible differences, such as the replacement of a certain number of cations, lattice flaws, etc., can be reflected, highly amplified, in their magnetic characteristics. The following considerations are general and limited to some aspects, more relevant for geological applications.

2.3.1
Magnetic Susceptibility

Magnetic susceptibility is the first parameter to consider: it enables to classify minerals in the three categories of substances, dia-, para-, and ferromagnetic, and thus to make a first evaluation of the magnetic properties of rocks. Moreover, susceptibility is the parameter that most influences the anomalies caused by rocky bodies to the Earth's magnetic field, and therefore it is extremely important in interpreting magnetic surveys (Chap. 3) for prospection and tectonic-structural studies. A comparison between minerals' susceptibility values provided in different texts may cause some confusion, because the values are often quite different. This is on the one hand due to the fact that minerals have a certain variability of their own, on the other hand due to a multiplicity of factors that influence the measurements, such as: synthetic or natural mineral, measurement on single crystal or on powdered material, powder grain size, anisotropy, etc. Our greatest interest is in classifying minerals, while the numeric values, compiled from various sources (Table 2.1), are meant to provide an idea of the order of magnitude and of variability.

- *Diamagnetic minerals*. To this category belong, in addition to ice, some of the most common minerals, in particular calcite and dolomite among carbonates, quartz and feldspars among silicates. Their susceptibility is in the order of -10 μSI.
- *Paramagnetic minerals*. These include many silicates, principal constituents of rocks, in particular pyroxenes, amphiboles, micas, clay-minerals. Their susceptibility is in the order of $100-5\,000$ μSI and it may have rather wide variations, also linked to the variability of Fe cations within a same mineral. For example, olivine can have any intermediate value between the ones corresponding to its two end-members: -13 μSI for forsterite (Mg_2SiO_4), diamagnetic, and $5\,000$ μSI for fayalite (Fe_2SiO_4), paramagnetic.

Table 2.1. Susceptibility of some common minerals

Mineral	κ (μSI)[a]
Diamagnetic	
Dolomite	–40
Calcite	–15
Quartz	–15
K-feldspars	–15
Gypsum	–15
Ice	–10
Paramagnetic	
Clynopyroxenes	20 – 600
Orthopyroxenes	1 000 – 3 000
Amphiboles	100 – 1 000
Biotite	800 – 3 000
Muscovite	40 – 700
Chlorite	70 – 1 550
Olivine	–13 – 5 000
Garnets	500 – 6 000
Ilmenite	300 – 3 500
Ferromagnetic	
Goethite	2 000
Hematite	1 000–5 × 10^4
Pyrrhotite	5 × 10^4 – 3 × 10^5
Magnetite	10^6–10^7

[a] 1 μSI = 10^{-6} SI.

- *Ferromagnetic minerals.* The most important are the Fe oxides and some Fe sulfides. Susceptibility is very variable: hematite (αFe_2O_3) can have values of just 1 000 μSI and magnetite (Fe_3O_4) can reach 10^7 μSI. Thus, the degree of oxidation has a fundamental role. Native Fe and Fe-Ni alloys are of interest only in specialist studies (meteorites, mantle nodules). Ferromagnetic minerals are not numerous and are mostly present in rocks in small quantities, as accessory minerals (<5%). Nevertheless, their extremely high susceptibility causes a content as low as 0.1% (Fig. 2.20) to mask the magnetic effects of the other minerals.

Fig. 2.20. Magnetic susceptibility of rocks versus the content of the main ferro- and paramagnetic minerals. Magnetite contents exceeding 0.1% mask the effect of all other minerals (from Hrouda and Kahan 1991)

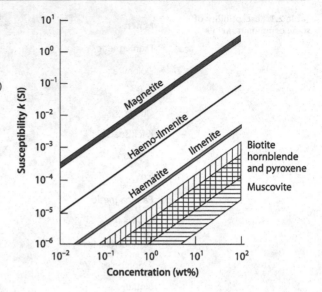

From the magnetic viewpoint, a rock can be considered to be formed by a dia- or paramagnetic matrix, within which are dispersed ferromagnetic grains which do not interfere with each other. The magnetic interactions between individual grains are effective over distances up to a few times their diameter, and the low ferromagnetic content means that average inter-granular distances are far greater than grain dimensions. This approximation is almost always valid, but it can become insufficient when the processes that formed the rock led to highly heterogeneous grain distribution.

2.3.2
Fe-Ti Oxides

The most important and widespread ferromagnetic minerals are Fe-Ti oxides, whose composition is schematically illustrated by the ternary system rutile (TiO_2) – wüstite (FeO) – hematite (Fe_2O_3) (Fig. 2.21). The terms rich in Fe are located near the base of the triangle, the ones richer in Ti towards the vertex. Fe^{2+} ions prevail to the left, Fe^{3+} ions to the right, which implies that the degree of oxidation increases from left to right. Natural Fe-Ti oxides form three isomorphic series: titanomagnetites and titanohematites (ferromagnetic *s.l.*), and pseudobrookites (paramagnetic). Commonly, natural minerals are in a higher oxidation state than in the theoretical formula and therefore they are shifted to the right along the dashed lines of the diagram (Fig. 2.21).

Titanomagnetites. These are the terms of the solid solution magnetite (Fe_3O_4) – ulvospinel (Fe_2TiO_4), whose general formula is $Fe_{3-x}Ti_xO_4$. They crystallize in the cubic system and their structure is that of spinels. The elementary cell contains 32 O^{2-} anions and 24 cations: 16 Fe^{3+} and 8 Fe^{2+} in magnetite, 16 Fe^{2+} and 8 Ti^{4+} in ulvospinel. The cations are distributed in two sub-lattices, A and B, to which respectively 8 and 16 cations belong per elementary cell. Within each sub-lattice the magnetic moments

Fig. 2.21. Fe-oxides in the rutile-wüstite-hematite ternary system. The *dashed lines* indicate an increasing degree of oxidation

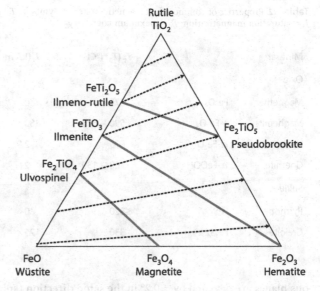

of the electrons associated with the cations are coupled parallel, but the two sub-lattices are mutually antiparallel. Consequently, magnetite is ferrimagnetic, because the sub-lattice B (8 Fe^{3+} cations + 8 Fe^{2+} cations) prevails on the sub-lattice A (8 Fe^{3+} cations), whereas ulvospinel is antiferromagnetic, because both sub-lattices A and B contain 8 Fe^{2+} cations. Interaction forces are reduced as Ti content increases and magnetic properties vary considerably. The Curie temperature of magnetite is 575 °C (Table 2.2), while that of the intermediate terms decreases almost linearly as Ti content increases, until reaching –153 °C, which is the Néel temperature of ulvospinel. Saturation magnetization J_s decreases from 480 to 0 kA m^{-1}. Most magnetic properties depend on the state of the grain: for example, the saturation remanence and the coercivity of remanence are greater for SD grains, smaller for MD grains.

Titanomaghemites. These are oxides characterized by numerous cation vacancies: they retain the spinel structure, but their chemical formula is that of titanohematites. Maghemite (γFe_2O_3) has magnetic properties that are similar to, though less pronounced than, those of magnetite, e.g. $J_s = 380$ kA m^{-1}. Its Curie point is 590 °C ≤ T_c ≤ 675 °C, variable as a result of the presence of impurities or of the different arrangement of the vacancies. Titanomaghemites are metastable; at high temperatures they undergo an irreversible transformation into titanohematites, with a drastic drop in the values of their magnetic properties.

Titanohematites. These are the terms of the solid solution hematite (αFe_2O_3) – ilmenite (FeTiO$_3$), whose general formula is $Fe_{2-x}Ti_xO_3$. They crystallize in the rhombohedral system and their structure is that of corundum. In hematite, the Fe^{3+} anions occupy basal planes, alternating with basals that contain the O^{2-} ions. Within each plane, the atomic magnetic moments are coupled parallel to each other, and between contiguous planes the coupling is not exactly antiparallel. The spins of the cations of two contigu-

Table 2.2. Properties of common ferromagnetic minerals Symbols: $T_c(T_n)$ = Curie (Néel) temperature; J_s = saturation magnetization; B_c = maximum coercivity

Mineral		$T_c (T_n)$ (°C)	J_s (kA m^{-1})	B_c (T)
Oxides				
Magnetite	Fe_3O_4	575	480	0.3
Maghemite	γ-Fe_2O_3	590 – 675	380	0.3
Hematite	α-Fe_2O_3	675	2.5	1–5
Goethite	α-FeOOH	60 – 130	2	>5
Sulfides				
Pyrrhotite	FeS_{1+x}	≤320	80	0.5–1
Greigite	Fe_3S_4	330	125	0.03

ous planes are deviated by ≈0.2° in the same direction (spin canting) inside the basal plane and they yield a small resultant, which results in a canted antiferromagnetic behavior. On the other hand, the coupling between the planes of Fe^{2+} cations of ilmenite is exactly antiparallel and hence its behavior (at very low temperatures) is antiferro-magnetic.

In this case, too, as the Ti content increases the Néel temperature decreases, drop-ping from 675 °C for hematite (Table 2.2) to –218 °C for ilmenite. Saturation magneti-zation J_s = 2.4 kA m^{-1} is smaller by two orders of magnitude than that of magnetite, while the coercivity of remanence is very high.

Fe oxyhydroxides. Goethite (αFeOOH) crystallizes in the orthorhombic system and it is antiferromagnetic, but as a result of anion vacancies it can have uncompensated spin moments which produce a weak magnetization. Its Néel temperature is ≈120 °C and it can decrease in the presence of impurities in the lattice. By dehydration in many natu-ral environments or heating in the laboratory, at ≈300–400 °C it transforms into hematite. It is characterized by a particularly high coercivity.

2.3.3
Fe Sulfides

The most widely known and common Fe sulfide, pyrite FeS_2, is paramagnetic, while pyrrhotite FeS_{1+x} and greigite Fe_3S_4 are ferromagnetic *s.l.* The magnetic properties of pyrrhotite depend on the value of x; for $0.11 < x < 0.14$, in the monoclinic structure there are two sub-lattices of coupled antiparallel Fe cations present, which because of vacancies have different magnetization. Hence, pyrrhotite is ferrimagnetic, and its Curie point is also variable as a function of x, in any case lower than 320 °C (Table 2.2). In metamorphic environments, pyrrhotite is often nickeliferous and the presence of Ni cations reduce the Curie point. Greigite has the same structure as magnetite, it is ferrimagnetic with Curie point near 330 °C. The J_s of these sulfides is in the order of 100 kA m^{-1}.

2.4 Ferromagnetic Minerals in Rocks

Most of the geological information that can be obtained from the magnetic study of a rock is contained in the small or even minimal fraction of ferromagnetic minerals it contains. Without getting into the complex field of magnetic mineralogy and petrology, let us briefly review, in this section, which ferromagnetic minerals are more common in the various types of rocks.

2.4.1 Igneous Rocks

In this case the two fundamental factors are the chemical composition of the magma, which controls the type of oxide that crystallizes, and the emplacement and cooling history, which control the chemical, mineralogic and thermal evolution of the crystals down to ambient temperature. Two guiding criteria for orientation (with all due care; to avoid excessive verbiage, in the following we will omit expressions like "usually", "in general", "often", ...) are therefore:

- *mafic/felsic rocks*. Mafic rocks have a relatively high Fe-Ti oxides content, up to 5%, and their Ti content is high: titanohematites have a composition close to ilmenite, whereas titanomagnetites have a highly variable Ti content, which can reach values up to $x \approx 0.6$. The Fe-Ti oxides content of felsic rocks is less than 1%, with lower Ti content: titanomagnetites have a composition close to magnetite and hematite may occur in addition to ilmenite.
- *effusive/intrusive rocks*. Effusive rocks cool rapidly and the dimensions of the crystals of Fe-Ti oxides are small, from a fraction of less than 1 μm to 0.1–0.01 μm: PSD grains are therefore common. During crystallization and cooling, physicochemical conditions are highly variable and primary oxides may undergo radical transformations, simultaneous or just subsequent to emplacement. Intrusive rocks have very slow cooling, especially if intruded in the lower crust, and crystal dimensions are in the order of 10–100 μm. MD grains prevail and crystals have the time to re-balance as physicochemical conditions change, with consequent advanced solid exsolution phenomena.

The history of the crystals of Fe-Ti oxides in igneous rocks can be quite eventful. They start to crystallize at temperatures exceeding 1 000 °C, for which a complete solid solution is possible within each of the series of titanomagnetites and titanohematites. At lower temperatures, starting from 700 °C, miscibility is no longer complete and the diffusion of Fe, Ti cations within a single grain leads to exsolution, with the formation of smaller regions, each characterized by a different content of the two cations. The result can be, for example, that an original MD titanomagnetite grain with intermediate composition is subdivided into smaller regions, partly with a composition closer to that of ulvospinel, and hence paramagnetic, partly with a composition closer to that of magnetite, hence ferrimagnetic in (probable) PSD state. In principle, solid exsolution is more pronounced in intrusive rocks, whose cooling is slower.

In effusive rocks, there are other, far faster transformations, linked primarily to oxidation. In sub-aerial rocks, lavas and pyroclastics, the high oxygen fugacity entails deuteric oxidation phenomena, which occur at temperatures in the order of 700–600 °C, thus mostly above the Curie temperature of Fe-Ti oxides. As a consequence, exsolution and intergrowth occur, which tend to follow the dashed lines in Fig. 2.21. In the case of an original grain of titanomagnetite, they lead first to the formation of alternating lamellae of titanomagnetite and ilmenite and then, as oxidation progresses, pseudo-brookite, rutile and hematite may be formed. As in the previous case, such processes entail radical changes in magnetic properties, such as pronounced reduction in magnetization if oxidation is very marked. In the case of submarine basalts, oxidation, caused by the presence of water and by heat occurs at low temperature, $T < 250$ °C. In this case, titanomagnetite is transformed into titanomaghemite: the transformation can be complete in grains with smaller dimensions, partial and limited to the outermost part in larger grains. Curie temperature increases with the degree of maghemitization and this process entails a stabilization of basalts' NRM.

These considerations allow us to have an idea of the magnetic characteristics of igneous rocks, always keeping in mind that the number of degrees of freedom is so high that this idea must always be substantiated by direct measurements on samples.

MORB basalts are characterized by titanomagnetite with high Ti content, up to $x \approx 0.6$, and hence Curie point <200 °C. Crystals have very small dimensions, due to rapid cooling. This entails very high values of J_r and of the Königsberger ratio, which may reach $Q = 100$. Maghemitization tends to cause J_r to be reduced, but simultaneously to become more stable, since both the Curie point and coercivity increase as the process evolves.

Sub-aerial basaltic and andesitic rocks generally have titanomagnetites that are definitely richer in Fe, with Curie point >500 °C. The values of κ and J_r are lower than MORB and $1 < Q < 10$. Sub-aerial emplacement entails a strong de-gassing and different cooling rates between the top and the inner part of a flow. The dimensions and oxidation of the grains, and hence their magnetic properties, may vary systematically according to the position within the flow, especially if the flow is thick.

Intrusive rocks of the gabbroic type, especially in continental crust, are characterized by Ti-poor titanomagnetites, close to the magnetite end-member. The slow cooling favors growth in grain size on the one hand, the occurrence of solid exsolution on the other. In any case, a considerable fraction of grains is MD, with consequent high values of κ and low values of J_r, with $0.1 < Q < 1$. A peculiarity of these rocks is the possible presence in many minerals, above all plagioclases and pyroxenes, of magnetite inclusions with dimensions <1 μm, hence in SD or PSD state. These inclusions do not radically change the values of κ and J_r, since their volume is negligible, but they give a highly stable NRM fraction to the rock. Hypoabyssal rocks, such as sills and dikes, have intermediate properties, obviously also as a function of their thickness. NRM tends to be very stable, thanks to reduced grain size and slightly longer cooling, which favors solid exsolution. Many Archaeozoic paleomagnetic data derive from rocks of this type.

Felsic rocks, as mentioned above, mainly contain small amounts of Ti-poor titanomagnetite. Values of κ and J_r can be very low, in effusive rocks also as a result of the more or less accentuated deuteric oxidation of titanomagnetite. In pyroclastic rocks, minute inclusions of titanomagnetite within vitreous particles may escape oxidation and give the rock a very stable NRM fraction. Granites have very diverse magnetic characteristics, depending on their origin. Those of the S-type (Chappel and White classification) contain mostly ilmenite, those of the I-type magnetite. Consequently, susceptibility is usually very low ($\kappa < 400$–600 µSI) in those of the S-type, higher ($3\,000$–$10\,000$ µSI) in those of the I-type. The characteristics of the NRM are highly variable according to the prevailing size of the grains and the cooling history, which in turn mainly depends on the depth of the intrusion.

2.4.2
Sedimentary Rocks

The ferromagnetic minerals of sedimentary rocks can be detrital, authigenic and diagenetic. The most common detrital minerals are titanomagnetites and titanohematites and their abundance in the rock depends in the first place on their abundance in the parent rocks. Thus, silt and sandstone have extremely variable values of κ and J_r: very low, if they derive from the disintegration of granitic rocks, higher if they are associated to basaltic volcanoes, very high if produced by the dismantling of serpentinite massifs, whose rocks can have particularly high magnetite content.

The distance between the site of origin and the site of deposition has an important role on the volume percentage of Fe-Ti oxides in the sediment. Their grains have densities in the order of $5\,000$ kg m^{-3} and therefore they tend to be deposited before those of siliceous minerals. Nonetheless, very small grains can be transported for thousands of kilometers, as in the case of the tephra produced by volcanic eruptions.

The ferromagnetic minerals of an unconsolidated sediment are in a precarious situation, because they are very sensitive to the presence of water and to the oxy-reduction environment. Thus, a detrital mineral can be completely transformed, and authigenic minerals can be formed which are not necessarily stable and which therefore may evolve over time. The subsequent diagenetic processes, with fluid circulation, pressure and temperature increase as a result of burial, can further modify the magnetic characteristics of the sedimentary rock. Let us review some possible processes. Red beds are typically formed in continental environment, in highly oxygenated waters: their intense coloring is caused by a hematitic pigment that pervades the rock and generally leads to a highly stable J_r. Black and greenish clays are formed in highly reducing environments, such as the bottom of marine and lacustrine basins with still waters poor in oxygen and rich in organic matter. These rocks typically contain sulfides. The Fe/S ratio changes as sulfurization advances, so from an initial precipitate whose approximate formula is $FeS_{0.9}$, the sequence is pyrrhotite (FeS_{1+x}) $\rightarrow$ greigite (Fe_3S_4) $\rightarrow$ pyrite (FeS_2). Depending on the available quantity of sulfur and on time, the sulfide in the rock thus has very different magnetic properties. Pyrite is very common, but it is not ferromagnetic. The other two sulfides are ferrimagnetic, but they tend to be transformed into oxides as diagenesis proceeds. Greigite has been identified mainly in Pliocene-Quaternary lacustrine and marine sediments. An additional complication derives from the fact that sedimentation conditions may change over time and the rock may be formed by layers with different magnetic characteristics. For example, variations in the detrital contribution and in oxygen content are typical of the delta environment, which can vary from lagoon (still water, high content of

organic matter → prevalence of sulfides) to canal (running water, high detrital content →
prevalence of oxides).

Carbonate rocks have an extremely low content of ferromagnetic minerals and they
are those with the lowest values of κ and J_r. The most widespread minerals are Ti-
magnetite of detrital origin and Ti-hematite, which probably represents its diagenetic
transformation, especially in the form of pigment which gives a reddish color to the
rock. Under particular conditions, goethite may be formed and preserved over time.
Deep ocean sediments have a highly complex magnetic mineralogy, because many
factors come into play, as detrital contribution, sedimentation rate, magnetosomes of
biologic origin, content of organic matter, dissolution linked to depth.

In soils, primary ferromagnetic minerals of detrital origin derive from the disintegration of the bedrock and they reflect its mineralogy. They are relatively abundant
in soils deriving from volcanic bedrocks (magnetite) or red sandstones (hematite).
Secondary minerals are formed through complex chemical and biological processes,
which also depend on climate and the soil pH, humidity and organic matter content.
These processes operate not only on primary ferromagnetic minerals, but also on the
elementary iron contained in many silicates. Depending on the parent material, the
physicochemical conditions and the pedogenetic processes, maghemite, goethite, hematite or magnetite can be formed.

2.4.3
Metamorphic Rocks

In this case, the factors at play are: the chemical-mineralogical composition of the
protolith, the metamorphic degree (pressure and temperature, P and T), the presence
and type of fluids, the tectonic stress, the time sequence of metamorphism, i.e. how P
and T conditions vary over time and when the new ferromagnetic mineral starts to
crystallize. Some common phenomena are the formation of chlorite at the expense of
Fe-oxides in low-grade metamorphic facies, the formation of sulfides in slates and in
high-grade rocks, and the formation of magnetite in the transition from amphibolitic
to granulitic facies. Ti-hematites can be formed and preserved in just about all metamorphic facies. More than general guidelines, case histories can be presented, but they
cannot always be extrapolated to other similar situations. There are numerous factors
at play and two protoliths with the same magnetic characteristics may give rise to two
different metamorphic rocks, with highly different magnetic characteristics.

However, some large-scale metamorphic processes are found in similar situations,
at least from the quality viewpoint, while the intensity of the phenomena can be highly
variable. One case is given by the formation of serpentinites by alteration of peridot-
ites; hydration of olivine at high temperature can take place according to the reaction:

$$2\ Mg_2SiO_4 + 3\ H_2O = Mg_3Si_2O_5(OH)_4 + Mg(OH)_2$$

forsterite   serpentine   brucite

This process, originally postulated by Hess to explain the formation of Layer 3 of
the oceanic crust, is particularly developed in the serpentinites included in the oro-

genic belts and deriving from tectono-metamorphic processes. Olivine contains a certain quantity of iron, which leads to the formation of magnetite, according to a reaction of the following kind:

$$6 \ Fe_2SiO_4 + 7 \ H_2O = 3 \ Fe_3Si_2O_5(OH)_4 + Fe_3O_4 + H_2$$

in olivine   *in serpentine*   *magnetite*

In the case of the Lanzo Massif (western Alps), the volume content of magnetite increases by a factor $\approx$20–30, as indicated by the magnetic susceptibility values: in the order of 1 000–3 000 µSI in the rocks of the peridotitic core, 60 000–80 000 µSI in fully serpentinized peripheral bands.

Another case, recently studied both in the western Alps and in the Himalayas, is observed in marly limestone sequences which contain detrital magnetite, when subjected to low-grade metamorphism at temperatures in the order of 300–350 °C. If a sulfur-rich fluid phase, which can derive both from gypsum intercalations and from organic matter maturing processes, circulates during metamorphism, pyrrhotite can form especially in black shales, according to a reaction of the following kind:

$$7 \ Fe_3O_4 + 12 \ S_2 = 3 \ Fe_7S_8 + 14 \ O_2$$

magnetite   *pyrrhotite*

Suggested Readings and Sources of Figures

Books

Butler RF (1992) Paleomagnetism: Magnetic domains to geological terranes. Blackwell Scientific Publications, Oxford, UK, 319 pp
Collinson DW (1983) Methods in rock magnetism and palaeomagnetism. Chapman and Hall, London, 503 pp
Dunlop DJ, Özdemir Ö (1997) Rock magnetism. Fundamentals and frontiers. Cambridge University Press, Cambridge, UK, 573 pp
Lindsley DH (ed) (1991) Oxide minerals: Petrologic and magnetic significance. Mineralogical Society of America, Washington D.C., USA (Reviews in Mineralogy, vol 25, 509 pp)
Lowrie W (1997) Fundamentals of geophysics. Cambridge University Press, Cambridge, UK, 354 pp
Merrill RT, McElhinny MW, McFadden PL (1996) The magnetic field of the Earth: Paleomagnetism, the core and the deep mantle. Academic Press, San Diego, California, 531 pp
O'Reilly W (1984) Rock and mineral magnetism. Blackie, Glasgow, UK, 220 pp
Stacey FD, Banerjee SK (1974) The physical principles of rock magnetism. Elsevier, Amsterdam, 195 pp
Tarling DH (1983) Palaeomagnetism. Chapman and Hall, London, 379 pp

Articles

Butler RF, Banerjee SK (1975) Theoretical single-domain grain-size range in magnetite and titanomagnetite. J Geophys Res 80:4049–4058
Hrouda F, Kahan S (1991) The magnetic fabric relationship between sedimentary and basement nappes in High Tatra Mountains, N. Slovakia. J Struct Geol 13:431–442
Pullaiah GE, Irving E, Buchan L, Dunlop DJ (1975) Magnetization changes caused by burial and uplift. Earth Planet Sc Lett 28:133–143

Chapter 3

Magnetic Prospecting

In the preceding chapters we have seen that the Earth's magnetic field is of internal origin to the Earth and can be considered, for about 90% of its magnitude, as that generated by a centered magnetic dipole. In the magnetic potential spherical harmonic analysis, the dipolar approximation corresponds to degree $n = 1$, with a representative magnetic dipole moment M of about 8×10^{22} A m^2. At a given Earth's point, the expression 'non-dipolar magnetic field' or generally 'magnetic anomaly with respect to the dipole', is normally referred to that part of the Earth's magnetic field that is left once the dipolar part, analytically expressible, is subtracted. This magnetic anomaly field, can be subdivided into two classes according to the anomaly spatial scale extension.

On the Earth's surface the anomaly field with respect to the ideal dipole shows a few, 8–10 in number, very large in spatial scale, about thousand km ($\lambda \sim 10^3$ km), anomalies that form the so-called *regional* anomalies. A second group of magnetic anomalies is formed by the so-called *crustal* anomalies and refers to those anomalies having spatial scales of a few km, or at most tens of km ($\lambda \sim 1$–10^2 km). This first approach subdivision follows the three broad domains structure typical of the Earth's magnetic field power spectrum (Sect. 1.2.6). When produced from satellite data, in fact, the power spectrum shows: the dipolar contribution ($n = 1$), a second domain, that refers to the non-dipolar contribution ($n = 2$ to 12–13) and a third domain, ($n > 12$–13) in which the crustal origin magnetic field is strongly represented (Fig. 1.16).

The dipole field, together with regional anomalies, is the part of the Earth's magnetic field having the deepest origin inside the Earth, and is also called, globally, the *main field*; this field is generated by an electrical current system flowing in the fluid part of the Earth's core, its origin then limited in extension by the core-mantle boundary. The part in the power spectrum for $n > 12$–13 has a shallower origin instead, being mainly the result of the field generated by the Earth's crust. The Earth's crust contributes to the Earth's magnetic field in two ways: (*a*) the remanent magnetization of crustal rocks, below the Curie temperature; (*b*) the magnetization induced in the crustal rocks by the main magnetic field. For this reason, the knowledge of the field originating in the Earth's core is central to study the dynamics of the core itself, but it is also fundamental to determine the part of the Earth's field of crustal origin. In fact the main field represents the fraction of the Earth's field that must be removed from that observed, for example on the Earth's surface, in order to isolate the crustal contribution, also called for this reason the *residual crustal field*.

Magnetometric measurements carried out on ships, helicopters, planes and also satellites, in low orbital altitudes, provide then information not only concerning the main Earth's magnetic field, but are also used to obtain information about the morphological configuration of the crustal origin field.

3.1
Instruments and Surveying Procedures

Some magnetic measurements are taken by means of portable magnetometers and are carried out along profiles or a series of parallel profiles in order to obtain a grid of data points. This is a broadly used method in magnetic surveys since it is finalized to the collection of a regular data set and allows to carry out measurements reasonably near to the anomaly sources. In very early magnetic surveys the so-called magnetic variometers, consisting of bar magnets suspended in the Earth's magnetic field, such as magnetic balances or dip needles, were frequently used. These instruments were often able to sense the Earth's magnetic field in all its components and allowed the complete characterization of the crustal magnetic anomaly; measurements were however very time consuming and the output could not easily be converted into digital information. From the 1950s a large success had the fluxgate magnetometers that were able to measure horizontal and vertical Earth's magnetic field components, with a sensitivity of 1 nT or so. Fluxgates are however relative instruments that are quite temperature sensitive so requiring correction; for this reason they were later on less used and they have been superseded by more effective and practical instruments.

Among the modern instruments employed for magnetic field measurements in surveys, the most widely used equipment for many years was the proton precession magnetometer. The digital output, the non critical orientation of the sensor, the intrinsic absolute stability and speed of operation, are among the objective advantages of these frequently used instruments, which can reach a 0.1 nT sensitivity. However proton precession magnetometers can only give a series of discrete measurements because of the polarizing and relaxation time needed by the protons. This drawback is overcome if Overhauser magnetometers are used instead. In the last 10–20 years or so, however, optical pumping magnetometers came frequently into play; they have, in addition, exceeded in sensitivity (now easily 0.01 nT) and measurement rapidity the precession magnetometers.

The sensibly fast procedure employed in surveys undertaken by means of proton precession, or optical pumping magnetometers, as well as their digital output, is however counterbalanced by the fact that these instruments provide only the total field intensity. As is well known the real observed magnetic field is a vector quantity instead, and intensity measurements therefore provide only part of the information. This characteristic gives rise to peculiar anomaly signatures that will be discussed in what follows.

The most effective method to cover large land areas with magnetic measurements is by helicopter, or plane, as needed. Since the aircraft that carries the instrumentation is very distant from the source generating the anomaly, special instruments are needed. In fact the magnetometer carrier flies at high altitude with respect to the ground, and then magnetometers employed in these surveys will need to be very sensitive in order to detect all requested details that refer to sources distant from the sensor. Marine magnetometry has also been very frequently used in the last fifty years or so to cover the vast ocean areas of the world. As in the case of all fast transport surveys the instrument sensor must be towed in a housing for stabilization and at a distance sufficient to keep it away from the complicate magnetic field the ship or the aircraft have of their own.

In order to obtain accurate positioning, modern GPS techniques are frequently used during the survey. Especially on marine magnetometry magnetic techniques are also frequently undertaken in conjunction with other geophysical investigation techniques, like gravity surveying and seismic profiling. A typical flight pattern for an aeromagnetic survey taken over Ross Island in Antarctica, and a photograph of a helicopter borne survey in action, are reported in Figs. 3.1 and 3.2.

Many industrialized countries have already surveyed most of their territory. For example, as regards aeromagnetism, North America is covered, for more than 80% of its area. Europe has been completely surveyed on the ground and by aircraft, although still detailed surveys in areas of great importance are still needed for specific scientific investigations.

As discussed in Chap. 1, the current unit universally used to measure the Earth's magnetic field and also used for magnetic anomalies is nT. This unit represents one of the most important achievements in the definition of a geophysical parameter, and perhaps this success is due to the fact that the nanotesla is directly connected to the old unit "gamma" used for a long time for quantifying magnetic anomalies. One gamma is equivalent to 10^{-5} Gauss and 10^{-9} T, thus representing a bridge between cgs-emu and SI system, where nT is equivalent to the old "gamma" (Appendix).

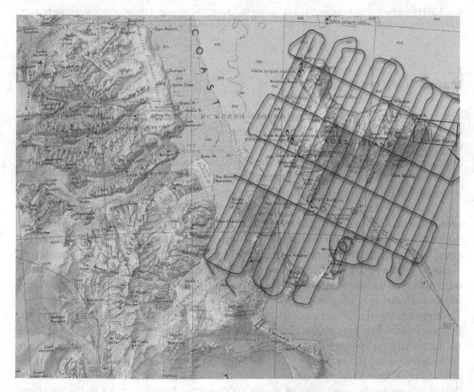

Fig. 3.1. A typical flight pattern for an aeromagnetic survey. The flight lines are superimposed on the geographic map of Ross Island, Antarctica (courtesy M. Chiappini)

Fig. 3.2. Photograph of an helicopter borne survey in action. The sensor of the magnetic instrumentation is kept at secure distance from the helicopter to avoid artificial magnetic effects (courtesy M. Chiappini)

3.2
Magnetic Anomalies

The crustal part of the Earth's magnetic field is generated in a very thin outer layer of the Earth with a maximum thickness of about 30–60 km, depending on tectonic setting, where ferromagnetic minerals can be found. Among these the most important are magnetite, titanomagnetite and hematite. Igneous and metamorphic rocks, below about 600 °C and composed of the above-mentioned minerals, are often characterized by a high magnetic susceptibility and therefore can be potential crustal magnetic field sources. Being the process that causes the rock magnetization related to the rock's geological history, the crustal origin field is important for example to obtain indirect information on geodynamic and tectonic structures and the thermal state of the Earth's crust. The identification of sedimentary basins, the determination of the boundaries between primary geological structures and orogenic belts, etc. lead to a detailed knowledge of the most superficial layers of our planet and can be obtained through the study of the crustal origin magnetic field.

Generally the size of a crustal anomaly, obtained by magnetic measurements, does never exceed a few percentage points, or in extreme cases 5–8% of the typical local field intensity, for the given latitude and longitude where the survey is undertaken. In these extreme cases it is the average remanent magnetization level, like in case of an igneous rock, that strongly modifies the field in its proximity, generating an anomaly that can even reach amplitudes of about 2 000–4 000 nT, with respect to the local expected main field value. Since all magnetic sources are essentially referable to magnetic dipoles their field decreases with distance as r^{-3}; moving away from the magnetization source, therefore the anomaly field tends to decrease rapidly in amplitude. In case of smaller remanent magnetization at a distance from the source, the field cancels out even more rapidly.

Since generally the Earth's mantle is considered to be non-magnetic, the top of the shallowest layer between crust-mantle transition and the hypothetical surface where the ferromagnetism-paramagnetism transition for magnetic minerals, occurs (the so-called Curie isotherm surface), is defined as "magnetic basement".

A very striking example of magnetic anomalies is given by the ocean magnetic anomalies. The surveying of the long linear sea-floor magnetic anomaly structures has been the milestone of plate tectonics revolutionizing the pre-existent geodynamic theories. At oceanic ridges the lithosphere is pulled apart allowing magma to rise to the surface leading to the formation of symmetrical alternating blocks of normal and reverse magnetized crust, due to the reversals of the Earth's magnetic field. The study of oceanic anomalies has allowed to determine the sea floor spreading velocity and the motion of transforming faults, thus allowing a quantification of these velocities.

3.2.1
Data Processing

Once a complete magnetic data set is obtained from measurements in the field, an initial data processing normally starts with the removal of extraneous data, such as spikes present in essentially all raw data. A careful visual examination of the whole data set is still a good procedure, however many different algorithms are now available for deleting and replacing spikes. If we exclude very limited area surveys, where location can be undertaken on the basis of visual marks or local grid systems, the location of survey points depends nowadays mainly on electronic positioning systems. GPS (satellite Global Positioning System) is the most widely absolute location system in use. Sometimes in airborne surveys film or videos are also taken during the flight for use as track recovery systems. All survey stations are afterwards numbered along with their coordinates and magnetic measurements are associated with geographic points.

The so obtained data set requires a processing procedure for obtaining an anomaly magnetic map. The quantitative definition of a crustal magnetic field, or anomalous field, at a given point P, is as follows

$$B_a(P) = B(P,t) - B_m(P,t) - B_e(P,t) \tag{3.1}$$

where B_a denotes the anomaly field, B the field measured at the given survey point P at the time t, B_m the main field at the same point P and time t, and B_e the external field at that point and time.

All possible magnetic field time variations of external origin and especially magnetic storms and magnetic diurnal variation must be removed from source data. Although the external part generally represents a small fraction of the Earth's magnetic field, its amplitude can be very variable and is influenced by latitude, solar cycle, season and so on (Sect. 1.3). Moreover, the external field time variation amplitude can be very large with respect to the precision needed in many magnetic surveys. For example at mid latitudes a strong magnetic storm can cause a variation of hundreds of nT in a few hours, and even in quiet days the diurnal variation can amount to tens of nT in a few hours. Removing the external magnetic field contribution requires data from a magnetic observatory, or at least from an independent simultaneously recording instrument that operates at a fixed nearby location for the duration of the survey.

For what concerns the computation of the main field B_m the reader can refer to Sect. 1.5 where the IGRF (International Geomagnetic Reference Field) was introduced; we will recall here only that the IGRF is an analytical expression that allows to compute for every point on the Earth the contribution due to the field generated by the Earth's core. For large extension surveys the IGRF allows the correct subtraction of horizontal main field gradients otherwise unknown. Sometimes the evaluation of the main field at a given location can be simplified. In case of small spatial extension surveys, for example, a local model of the Earth's magnetic field can be obtained as a mathematical representation by means of power series in latitude and longitude. This is sometimes called a *normal field* and allows to reconstruct the spatial and time variations of the Earth's magnetic field over the given area directly from a subset of survey measurements. In very limited areas, as for example in archaeological or environmental applications, the main field does not vary significantly in the measured area, so that even a single average value, representative of the main field for all the surveyed region, can be assigned to the all area.

In airborne surveys, especially, measurements along a series of tracks called tie lines, perpendicular to survey lines (generally at a wider spacing) are also performed (Fig. 3.1). These supplementary data are used as valuable data for the survey and also to compare data at the intersections with the aim to reduce or eliminate differences between the field values measured at the same geographic points. An empirical mathematical procedure, normally called leveling, consisting of a linear least squares fit is normally employed for the correction. When the data set has been processed an interpolation to a regular grid is often required before a magnetic anomaly profile or a contour map can be prepared.

3.2.2
Summary of Operations

Summarizing, a magnetic survey aimed at the determination of the crustal origin magnetic field, must involve the following procedure:

1. to carry out magnetic measurements on profiles or grids in the selected area;
2. to edit the data set for spikes and assign correct measurements location;
3. to subtract the magnetic field time variations from all measurements:
   a rapid time variations, from observatory or independent local temporal station recordings;
   b slow time variations, namely SV should be included in the main field only if the survey required long time for the execution or if measurements were made in different epochs and are finally collected in one survey epoch. In this case a nearby observatory absolute data is often necessary to reconstruct the secular variation in the given time window;
4. to subtract the field value due to the main field part, at all measurement points and leveling and gridding data if required.

As in the case of a gravity anomaly, that results as a superposition of a local contribution to the global Earth's gravity field, a magnetic anomaly results from the superposition of a local magnetic source to the main Earth's magnetic field. In the case of magnetism, however, the local contribution can be very complex and, as will be seen in what follows, can be due to a number of possible crustal sources. Moreover whereas the gravity field is by definition always vertical, everywhere, the shape of the Earth's magnetic field lines is more complex and, in addition, the geomagnetic field varies not only in amplitude but also in direction. The profile or contour lines of magnetic anomalies and the following interpretation of survey results depend largely on geographic parameters. In fact magnetic induction, due to the action of the main magnetic field on crustal rocks, depends on local field direction in space. Magnetic inclination, and also declination, are variable from place to place on Earth, so that obtained magnetic anomalies can vary significantly in shape especially as a function of latitude, since magnetic inclination is systematically dependent on latitude.

3.3
Significance of Magnetic Anomalies

The processing of a magnetic survey data set gives as output a profile along a given direction, or a map, on which the crustal magnetic anomalies are plotted. The case of a profile results in a x,y-diagram where the magnetic anomaly is plotted versus distance. Maps are generally provided as contour lines or, more frequently nowadays, as a color contour map. From these representations the most important and complex phase starts, that of interpretation. The experience of the interpreter, the knowledge of the surface geology, the tectonic structure and the survey dimensions, are important elements to take into consideration to obtain a reliable interpretation. In gravity anomaly studies, the factor determining the anomaly pattern in the geological context, is the density contrast between the source of the anomaly and the surrounding rocks ($\Delta\rho$); in magnetism the critical factor is the magnetization contrast between the source and the rocks in which it is included (ΔJ). The cases of gravity and magnetism could be considered very similar since their respective fields are very similar; for example they can both be expressed by means of a potential solution of Laplace's equations. However some relevant difference between the cases of gravity and magnetism emerge when we examine anomalies more carefully.

In the case of magnetism the presence of remanent magnetization can considerably modify an anomaly pattern. Some rocks in fact possess a remanent magnetization J_r which, added to that induced by the present main magnetic field J_i, provides a total magnetization. The remanent magnetization is generally very stable and independent from the present-day field, being the result of a process that had taken place at the time of the rock's formation and of all other possible events that the rock has undergone in its history. Sometimes the remanent magnetization can also considerably exceed in intensity the induced magnetization. Referring to Eq. 2.5

$$J_{tot} = J_i + J_r = \kappa H_E + J_r$$

the contribution of the induced and remanent magnetization to the total magnetization of a rock, J_{tot}, depends on the Königsberger ratio $Q = J_r / J_i$, which varies over three orders of magnitude depending on lithologies (Sect. 2.4). It can exceed 100 in mid ocean ridge basalts and be less than 0.1 in sedimentary rocks.

Therefore the total magnetization J_{tot} is a vector sum of the two magnetizations and can point to an arbitrary direction in space. Only if J_r is parallel or antiparallel to the present field, the resulting magnetization is the simple algebraic sum of the two magnetizations (see Sect. 4.5.4). Generally this condition is not frequently experimented (the stability of remanent magnetization is in paleomagnetism the fundamental assumption in plates movement reconstruction). Although rock magnetism allows to separate, in laboratory, the contribution of the induced magnetization from remanent magnetization, this separation is not possible in the field.

If the geophysical survey through magnetic methods is carried out in a region where Q is reasonably low (for instance often in case of continental crust) the magnetization can be considered as entirely induced by the present field, $H_e = F / \mu_0$, and so parallel to F and proportional to the magnetic susceptibility. If k_0 is the susceptibility of the surrounding rocks and k is the susceptibility of the source-body of the anomaly, the magnetization contrast determines the size of the anomaly:

$$\Delta J = (k - k_0)H_e \tag{3.2}$$

If the anomaly source is generated only by magnetic induction, a total magnetic field survey carried out at the magnetic poles, where the field is almost vertical, follows very closely the case of a gravity survey. Moreover if the body causing the anomaly is of vertical extension nearly infinite, so that only one polar extremity is near the ground, we will have a reasonable resemblance to a monopolar case, which is similar to the gravimetric one. If the body has finite dimensions, the source will have to be represented by a dipole and therefore the anomaly profile, even though similar to the monopolar case, must be mathematically defined in a different way. In fact in this case the magnetic field direction, produced by the body, will be concordant with the local field, above the anomaly source, but opposite at its edges, providing a positive anomaly only exactly above the body, but with negative values at its sides.

If the anomaly source is generated by a crustal body by induction and the total magnetic field survey is carried out in an equatorial zone, the anomaly profile differs considerably from the polar case previously mentioned. In fact in this case the body magnetization produces a field that is opposed to the inducing field above the body itself, with two positive lobes at the two sides. At intermediate latitude, where the main field makes an angle with the horizontal plane, we will find a still different situation. Generally in the northern hemisphere, we obtain profiles that show positive anomalies offset with respect to the center of the body towards the south: a minimum appears at the northern part and a maximum at south; the intensity of the positive and negative lobes depend on the local inclination value. In the southern hemisphere we obtain the opposite behavior.

Under conditions of almost zero local declination, the axis that connects the minimum and the maximum of the anomaly configuration is oriented along the north-south direction. If the declination differs from zero, its value influences the azimuth of the axis connecting the center of the positive anomaly with that of the negative one, causing an angle with respect to the north-south line. In Fig. 3.3 magnetic anomalies

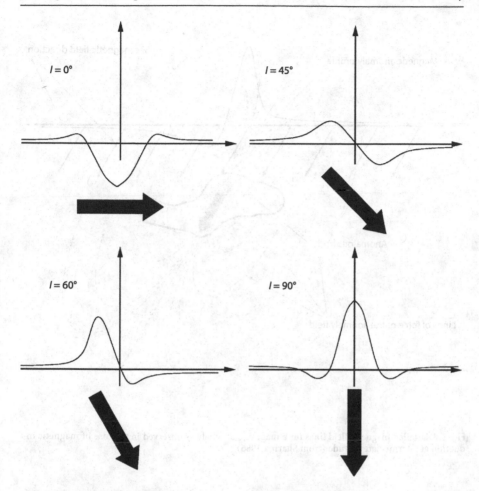

Fig. 3.3. Magnetic anomaly along a S-N profile for different inclination angles in the case of simple magnetic induction in the Earth's magnetic field (northern hemisphere). *Black arrow* represents the magnetic moment of the source-body

produced by induction at different latitudes represented by different inclination angles are reported. In Fig. 3.4 the case of an intermediate latitudes subsurface body magnetic anomaly, is drawn together with the lines of force of the induced magnetic field.

3.3.1
Forward Modeling

The magnetic response produced by a body of any given shape to a magnetic induction process, or the response of a vertically stratified ground to magnetic induction, are typical examples to which various authors have given mathematical solutions. Some high-quality scientific and commercial software, able to represent the anomaly field expected in these basic cases, is now available. Basically every source of unknown shape

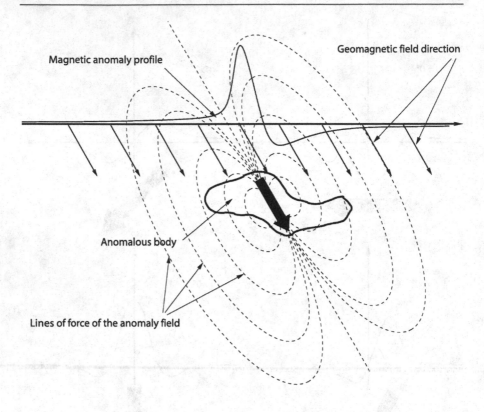

Fig. 3.4. Detailed magnetic field lines for a magnetic anomaly as observed in the case of magnetic induction at intermediate latitude (from Sharma 1986)

that causes a magnetic anomaly, can be represented by an assemblage of magnetic dipoles. We can therefore envisage any given body as being subdivided in a series of infinitesimal volume elements ($d\tau$) to which a magnetic dipole moment ($\mathbf{m} = J\,d\tau$) is associated.

We have already observed that Maxwell's equations, under ideal magnetostatic conditions, have allowed to obtain, for a scalar magnetic potential V, at any distance r' from a source with magnetization J, the equation

$$V(r') = -\frac{\mu_0}{4\pi} \int_\tau \frac{|\nabla \cdot \boldsymbol{J}(\boldsymbol{r})|}{|\boldsymbol{r} - \boldsymbol{r}'|} \, d\tau \qquad (3.3)$$

where τ denotes in this case the integration volume. Using the properties of vectors the above equation can be reformulated allowing the gradient operator to go outside the integral and rewriting in the following simpler way:

$$V(r') = -\frac{\mu_0}{4\pi} \nabla \cdot \int_\tau \frac{|J(r)|}{|r-r'|} d\tau \qquad (3.4)$$

From this general formula, given a particular magnetization distribution in an assigned volume, the magnetic potential produced by this known distribution and therefore the magnetic field associated with it, can be computed in the case of simple geometric source cases. From Eq. 3.4, the magnetic anomaly field component T projected along a generic direction f in space and t is the body magnetization direction, is:

$$\Delta T(r') = -\frac{\mu_0}{4\pi} \frac{\partial^2}{\partial f \partial t} \int_\tau \frac{|J(r)|}{|r-r'|} d\tau \qquad (3.5)$$

The direct comparison of measured anomalies, as obtained from a magnetic survey, with the anomaly distribution field obtained by a given model, is called *forward modeling*. The model image or profile can be compared with data obtained from the survey in order to reach, by subsequent approximations in the model parameters, the ideal model that gives the closest agreement with the observed ground data *(trial and error)*.

We will now only show here a simple modeling example in the case of the anomaly generated by a buried sphere magnetized by an external vertical magnetic field; in this case we will easily obtain the resulting anomalous field (Fig. 3.5). From potential theory it is possible to show that the magnetic field, generated by a uniformly magnetized sphere, is equivalent to that of a dipole moment M, placed at its center; using the gradient formulation:

$$V = \frac{\mu_0}{4\pi} M \cdot \nabla \left(\frac{1}{r}\right)$$

that is

$$V = \frac{\mu_0}{4\pi} \frac{M \cdot r}{r^3} = \frac{\mu_0}{4\pi} \frac{M \cos\theta}{r^2} \qquad (3.6)$$

If we consider a polar coordinate system and refer to the transverse, F_t, and the radial, F_r, components we obtain (Eq. 1.4)

$$F_r = -\frac{\partial V}{\partial r}; \quad F_t = -\frac{\partial V}{r\partial \theta}$$

In areas where the field is almost vertical, as can be assumed for $I > 60°$, we can obtain the F_z component directly. The magnetic effect on the vertical will be

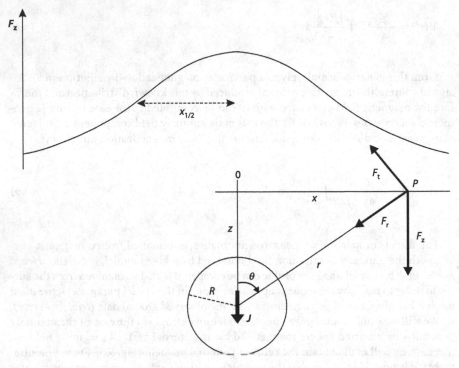

Fig. 3.5. Magnetic anomaly field due to a vertically magnetized sphere according to the geometry described in the text

$$F_z = \frac{\mu_0}{4\pi} \frac{2M\cos\theta}{r^3}\cos\theta = \frac{\mu_0}{4\pi}\frac{2M}{(x^2+z^2)^{3/2}}\left(\frac{z^2}{(x^2+z^2)}\right) \tag{3.7}$$

The spherical body formula defines an anomaly profile that can be plotted on a graph. This anomaly profile can be compared to real ground measurements; in case a similarity is found the model results can be varied by modifying the given parameters to best fit the real data, allowing in this way to infer the buried body parameters. In Fig. 3.5 the magnetic anomaly due to a vertically magnetized sphere, is represented for the case in which the vertical magnetic field is measured. In the analysis of subsurface structures by means of magnetic methods one of the most important and simple parameters to estimate is the depth of the anomalous structure. In case of simple geometrical body shapes, such as the above mentioned sphere, frequently so-called depth rules are available; the half width of the anomaly profile $x_{1/2}$ can be used as such a depth estimation.

Other particular cases that include prisms, cylinders, or bodies extending in space, for example in the horizontal direction to approximate cases of dikes, strikes or faults, can similarly be obtained.

In conclusion in the forward modeling approach the crustal field, under study, is obtained on the assumption that it is due to fields generated by known buried dipolar sources. As we have just noted, for example, a spherical magnetized body is always represented by means of a dipole placed at its center. The interpretation of magnetic surveys has relied for many years on forward modeling techniques, being this the only one used in practice. In the last 20 years or so the availability of portable and inexpensive computers has made forward modeling much easier also with the possibility of using interactive procedures involving computer graphic systems.

3.3.2
Inverse Modeling

Inversion techniques in the case of potential fields in geophysics, represent an interpretative tool that has had a very exciting development in the last few years. Substantially the term inversion means a mathematical technique in which the source model, that generates a given observed anomaly is reconstructed directly from the observed data. In contrast to what happens in the case of forward modeling, where a comparison with a preliminary model is needed and the final model is obtained by trial and error, inverse modeling outputs a model of the Earth only in terms of distribution of some physical properties, directly based only on observed data. Of course an inverse problem does not bring to a unique solution. Gauss's theorem, in fact, states that, in the case of potential fields, when a magnetic field is produced by a given body, a volume magnetization distribution (in the limiting case a central point magnetization) or a surface distribution of magnetization, are both possible equivalent solutions. Consequently in every inversion method we can resort to approaches that allow only to come close to the most reasonable solution for the given problem.

A real free inversion should take into consideration no restrictions at all, or only very few limited restrictions, as for instance an appropriate depth function for the inferred sources. Some inverse methods make use of an approach that takes into consideration probable limitations based on the knowledge of the geological structure under investigation; this is for example the case when geological evidence allows to infer a reasonable geometry for the source.

Several techniques of inversion in geophysics are currently in use; one of the best-known is the Parker method that is based on an iterative scheme involving Fourier transforms to find the distribution of the underlying magnetization. In this paragraph we will only give an overview when observed data can directly be related to the source parameters, as in the case of magnetic susceptibility or magnetization. This is the case when the observations and the source parameters can be represented by a linear equation system. We will then show an approach, that can be referred to as a matrix method, that allows a simple example data inversion. Modeling the anomaly field sources with some assistance from geological information available in the survey area simplifies this approach.

Magnetic data from a survey can be inverted by means of a procedure that starts by subdividing an ideal three dimensional body, underlying the surface where the survey was undertaken, in a series of m infinitesimal volume elements, for instance in the form of very small vertical rectangular section prisms, each of uniform magnetization J_i. Naturally in the three-dimensional volume the number of cells is in general much larger than the available data, say n measurements, making, therefore, the problem by definition under-determined. This situation appears realistic considering that a surface data distribution, resting on two dimensions (as is the case in a ground or aeromobile survey) cannot reasonably allow to obtain a full information on the underlying three dimensional volume. This is essentially referable to the impossibility to distinguish, starting from data outside a given surface, between a configuration of surface sources and a volume source configuration (Gauss's theorem).

We can then consider the case of a linear equation system where y_i are the anomaly field values on the surface and each cell in the model corresponds to a uniformely magnetized prism of assigned magnetization J_i. We can now call $x_{i,j}$ the matrix (also referred to as *kernel*) that represents the contribution to the magnetic anomaly in the i point given by the j prism. For N small prisms we will therefore have:

$$y_i = \sum_{i=1}^{N} x_{i,j} J_i \tag{3.8}$$

Consequently the inverse problem is reduced to the search for a solution of a system of equations with the aim to find the J_i values. Clearly, on given assumptions, it is theoretically impossible to invert the system directly, in fact only a so-called pseudo-inverse solution is possible; the Levemberg and Marquardt method, for example, constitutes an iterative numerical algorithm that allows to calculate a pseudo-inverse solution for the given system of equations.

This procedure may appear simple but we should remember that: (*a*) observed data are always contaminated by errors that will not help to reach a stable solution, and (*b*) the search for exact theoretical modeling for the prisms can make the linear equations system mismatched. In order to obtain a reliable solution to the inverse problem therefore some restrictions to overcome these obstacles are necessary. Notwithstanding that, in any case a purely good mathematical solution to the problem could not necessarily be a good physical solution.

To impose some pre-assigned condition to the modeling, is always a sensible procedure. For example using some previous independent knowledge of the subsurface volume, such as a given interval for the small prisms magnetizations (maximum and minimum values) and, moreover, reasonable limits to their depth distribution. Furthermore a reasonable solution would generally require the compactness of the source, in fact a genuine arrangement of the expected source as a whole is more realistic then a dispersion of single dipoles (trial called in many cases regularization). In addition we note the possibility of insertion of a 'weighting function', opportunely chosen, to drive the algorithm towards a sensible physical solution.

As previously mentioned, at present inverse methods are able to furnish only approximate expected models for the structure that causes the measured anomaly field. Being commercial softwares that realize inversion procedures available only in certain cases, inverse methods are generally at an experimental stage. Studies finalized to the search for the best weighting functions, general restrictions or new mathematical inversion techniques, are in progress, confirming that inverse modeling is a field in full development.

An example of the capability of inverse methods is given in Fig. 3.6 in three panels. In Fig. 3.6a an assemblage of magnetic prisms is represented schematically for an ideal subsurface body subdivided in a series of prisms; in Fig. 3.6b the surface anomaly produced by the synthetic body is shown as a three dimension map. In Fig. 3.6c the final result of an inversion procedure, as applied to the synthetic magnetic anomaly map data of Fig. 3.6b, is shown. The result of the inversion appears as slices of magnetization reported on a gray scale, at various depths resembling the synthetic body.

3.3.3
Spectral Characteristics

When a magnetic anomaly profile, or a magnetic contour map is obtained, the selection of the correct meaning to attribute to the single contribution intensities that appear in the plot, is frequently a difficult task. The ideal case of the buried sphere, previously introduced, is a simple case that immediately allows the selection of the possible source since the resulting magnetic anomaly profile is quite regular and intelligible. The real cases are normally quite different. In the presence of diverse sources of magnetization, or of a single body with varying magnetization, experimental anomaly profiles result in very irregular curves. In real cases an advantage can be obtained from an objective methodology that allows to evaluate quantitatively the degree of roughness or smoothness in the anomaly profile. This is possible with the use of Fourier transforms. This technique is normally used for interpretation in one or two spatial coordinates, according to the need, to quantify the degree of roughness in the profile by isolating, even if not individually but at least for groups, the responsible harmonic components.

If spectral analysis results are plotted with the amplitude of harmonic components as a function of the harmonic degree, generally on a semi-logarithmic scale, the anomaly is analyzed in terms of the effects produced by the single wavelength contributions. In this case it is often possible to isolate two or more linear sectors in the spectrum. Each sector, corresponding to a given wavelength interval, can be attributed to a given depth interval for the corresponding sources (or better interval of depths of the sources distribution).

A very useful technique in the analysis of the complete magnetic anomalies spectrum is denoted as 'upward continuation'. The purpose of this numerical technique is to obtain the anomaly information as the result of an ideal upward continuation, as if the anomalies were observed at a higher elevation than that at which measurements have been taken. The advantage is that often the anomaly produced by sources that are closest to the measurements level, are those that mostly influence the measurement, introducing a strong variability and bias in the observed field. The upward continuation allows a picture of the field at a greater altitude making the deepest sources to show up more clearly.

The numerical technique normally employed for upward continuation is generally based on the use of Fourier transforms; once the spectrum is obtained at the mea-

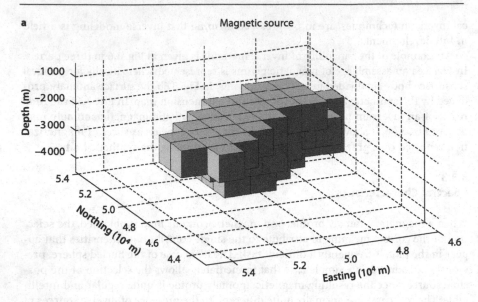

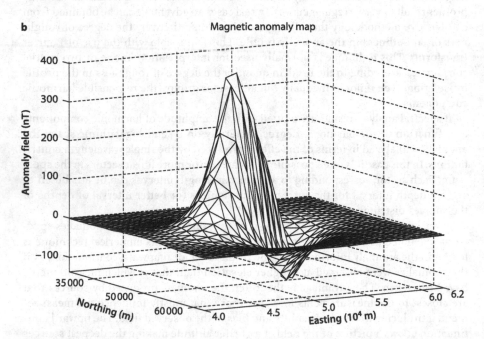

Fig. 3.6a,b. A magnetic anomaly inversion procedure capability; **a** an assemblage of magnetic prisms is represented schematically for an ideal subsurface body subdivided in a series of prisms; **b** the surface magnetic anomaly produced by the synthetic body represented in Fig. 3.6a is shown as a three dimensional map (courtesy A. Pignatelli)

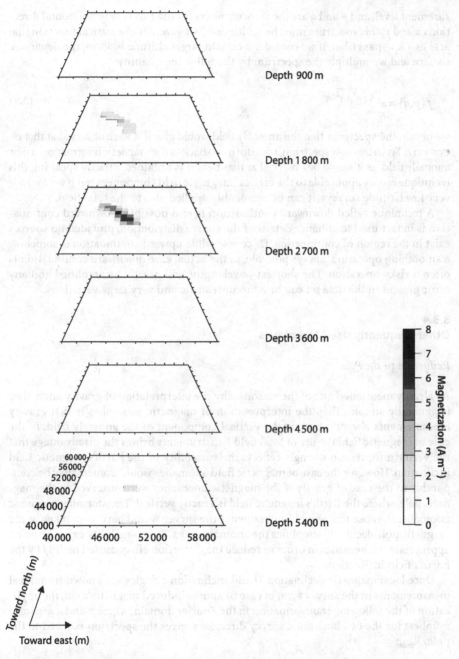

Fig. 3.6c. A magnetic anomaly inversion procedure capability; the result of a magnetic inversion on the anomaly map in Fig. 3.6b is reported as slices of underground magnetizations, on a gray scale, at various depths. This map can be compared to initial model in Fig. 3.6a (courtesy A. Pignatelli)

surement level, and p and q are the wave numbers for the two survey horizontal directions x and y, this spectrum must be multiplied by a given mathematical function that acts as a low-pass filter. If we consider a certain target altitude h, above the departure surface and we multiply the spectrum by the following quantity

$$f(p,q) = e^{-h\sqrt{p^2+q^2}} \tag{3.9}$$

we obtain the spectrum that the anomaly field would give if it were observed at that elevation h. From this new spectrum, transforming back to the magnetic information, a new anomaly field, as it would be observed at that height is obtained. Strictly speaking this technique is only applicable to the vertical magnetic field component but if we exclude very low latitude surveys, it can be reasonably applied also to the total field.

A technique called downward continuation is also possible. Downward continuation is in fact used to enhance details of the source distribution, provided no sources exist in the region of continuation. Of course while upward continuation is somehow a smoothing operation, always possible, in the actual case downward continuation is often a risky operation. The shortest wavelengths will greatly be amplified and any error present in the data set can bring to unrealistic and very large variations.

3.3.4
Other Frequently Used Techniques

Reduction to the Pole

As already mentioned one of the reasons why the interpretation of gravity anomalies is generally simpler than the interpretation of magnetic anomalies is that gravity measurements always represent the vertical component of the anomaly field. In the case of magnetic field the use of total field F instruments brings the disadvantage that the vector information strongly reflects the variability of the Earth's magnetic field inclination. However the case of magnetic field anomalies would immediately be comparable to the case of gravity if the magnetic anomalies were observed at the magnetic pole, where the Earth's magnetic field is exactly vertical. Transforming magnetic anomalies to those that would be obtained if the survey would have been taken at the magnetic pole decidedly simplifies the anomaly maps. This possibility exists by using appropriate mathematics in order to reduce the distortion effect caused mainly by the Earth's field inclination.

Once local magnetic declination D and inclination I angles are known from local measurements in the survey area, in case of simple induced magnetization, the application of the following transformation in the Fourier domain, where p and q are wave numbers for the two horizontal survey directions, gives the spectrum reduced to the pole, R_{pole}:

$$R_{pole} = \frac{p^2 + q^2}{i\cos I \cos D p + i \cos I \sin D q + \sin I \sqrt{p^2 + q^2}} \tag{3.10}$$

Generally only one declination and inclination value is considered for all the surveyed area. However particular codes are now available also for cases in which the declination and inclination angles would significantly vary in the surveyed area. In Fig. 3.7 an ideal case of reduction to the pole is shown to illustrate the magnetic anomaly pattern modification produced by this technique.

Magnetic Gradiometer Surveys

Sometimes magnetic surveys are undertaken by the use of magnetometers with a special sensor arrangement: two or more sensors are simultaneously operating so to give magnetic measurements that can be used to obtain the local magnetic field gradient. For example with two sensors placed at the vertical spacing of only 1 m, it is possible to get a very useful gradiometric information. This value, expressed generally in nT/m in fact furnishes a measure that is free from magnetic time variation effects that, as seen before, cause the need of a correction of the survey data; in the case of gradients this is not necessary. Magnetic gradiometric surveys allow to enhance some geometric characteristics of the source, for instance emphasizing anomalies from shallow sources and allowing an easier detection of causative body edges.

Analytical Signal Method

When the survey data are not taken with a gradiometer instrument similar advantages can be obtained with mathematical operations on ordinary magnetic anomaly data. The method of the analytical signal for example, is based on the use of the spatial derivatives of magnetic anomalies computed along three orthogonal directions. By this operation a total absolute value of the analytical signal of a given anomaly field can be obtained by summing the three square differentials. The analytical signal allows, for example, to enhance the edges of the geological structures in a magnetic anomalies cartography.

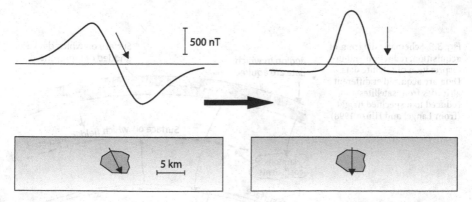

Fig. 3.7. Ideal case of reduction to the pole: a typical mid latitude magnetic anomaly pattern (*left*) is modified by this technique (*right*) (redrawn from Blakely 1996)

3.4
Satellite Magnetic Measurements

Artificial satellites have given a significant contribution to enrich our Earth's magnetic field knowledge. The importance of magnetic field measurements from satellite altitude is straightforward. Satellite data furnish in fact a global coverage of the Earth's magnetic field morphology in a short time compared to the Earth's magnetic field time variations, for example the secular variation. Global coverage satellite measurements are taken over a certain time period and at different altitudes, since the orbit is generally elliptical. They are then processed and reduced to the same epoch and at the same altitude. In Fig. 3.8 a representation of the actual situation in a satellite survey is given together with the indication of the interested regions. From the values thus obtained, a global map of the Earth's magnetic field is derived and then processed further according to the needs. Also in the case of satellite data the main challenge is to separate the Earth's core contribution from that of the lithosphere. Both fields contribute to the intermediate spherical harmonics of degree $12 \leq n \leq 16$ and a mutual contamination in the separation process is inevitable.

Satellite magnetic data can be used in a variety of magnetic field investigations. Models of the geodynamo can be studied from magnetic field satellite maps and also downward magnetic field continuation to the CMB can be carried out. Satellite measurements can also be used for studies of the lithospheric field, that contains a great deal of useful geological information. Moreover crustal field magnetic anomalies with wavelengths larger than 500 km can be studied mainly by satellite, since they are not obtainable simply pasting together magnetic surveys made on the Earth's surface. Only the satellites can offer this large scale magnetic perspective.

Some problems are evidently found also with satellite magnetic measurements. Satellites orbit at an average altitude of 300–800 km, where main-field values are somehow similar to those found on the Earth's surface but crustal anomalies intensity decrease considerably. Satellites orbit in a very active electromagnetic environment. Electric currents in fact circulate in the ionosphere especially at lower altitudes than the satellite orbit in particular the Sq currents and the equatorial electrojet (Sect. 1.3.4).

Fig. 3.8. Schematic diagram of acquisition, reduction and interpretation of satellite data. Data are acquired at different altitudes from satellites and reduced to a specified height (from Langel and Hinze 1998)

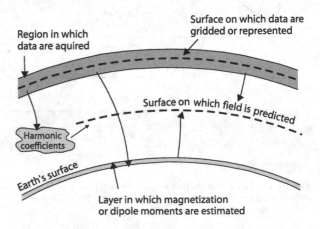

These currents generate curl-free magnetic fields whose source is internal to the orbit, just like those generated in the core and the lithosphere. This contamination can be identified and partly eliminated thanks to the fact that ionospheric currents follow the Sun's apparent rotation, so their effects vary with local time. These are reasons why satellite maps published by different authors often are broadly similar, but can differ in the case of individual anomalies.

The satellites OGO 2, 4 and 6 (also called POGO) were sent in space orbit between 1965 and 1971; they have conducted the first good quality measurements to enable the determination of the main, and crustal residual, Earth's magnetic field from space. A noteworthy success came after MAGSAT, the first satellite ever been made on purpose for the study of the Earth's magnetic field, was born. The value of the total magnetic field as well as of its magnetic components, was obtained. Launched in October 1979 this satellite has remained in a sun-synchronous orbit between altitudes of 352 and 578 km a.s.l., recording magnetic data until June 1980, when the satellite has burnt in its re-entry in the lower atmosphere. For several years MAGSAT data have been the best available satellite data for the study of the main as well as the crustal magnetic fields (Plate 1).

Despite its short life span and the high altitude orbit, MAGSAT has contributed remarkably to the progress in our knowledge about the lithosphere magnetization, showing that the weak lithospheric magnetic field is indeed discernible at the satellite altitude. Unfortunately because of the high level of noise introduced in the measurements by ionospheric currents at satellite altitude, and the eccentric orbit, field models deduced by various authors from the initial data, were not in good agreement among them. The general level of the intensity of the lithospheric magnetic field differs in fact, by a factor 2 between the various models, corresponding to a factor 4 in the difference between the power spectra.

After these initial successes, for about twenty years no more magnetic satellites were constructed. Fortunately new satellites were made recently specifically for magnetic measures in order to furnish new data. In the month of February 1999, the satellite ØRSTED was lunched. This satellite orbits at an elevation between 620 and 850 km, with an elliptic inclination of 96.62°, therefore approximately in polar orbit. ØRSTED is equipped with Overhauser and tri-axial fluxgate magnetometers for measuring the total intensity and the components, respectively. In July 2000, launched in an almost polar circular orbit, with an initial altitude of 455 km, eccentricity approximately 0.001, magnetic instrumentation of high precision, and with an expected life of about 5 years, the new satellite denominated CHAMP operations started.

3.5
Applications of Magnetic Anomalies

As we have seen in this chapter exploration through magnetic techniques is often finalized to the determination of the crustal magnetic field. It therefore has implications in tectonics, mining and oil exploration, and is also very important for a better understanding of the nature and evolution of all the lithosphere (Chap. 6). Magnetic surveys have supported petrologic studies, allowing the identification of minerals, oil, archaeological artefacts and are recently often used also in environmental geophysics studies.

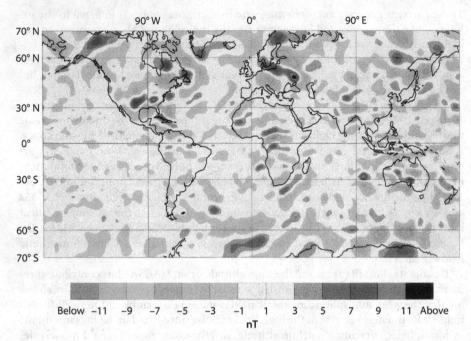

Plate 1. Map of the total intensity anomaly of the Earth's magnetic field at 400 km altitude derived from satellite data (POGO and MAGSAT) (from Arkani-Hamed et al. 1994)

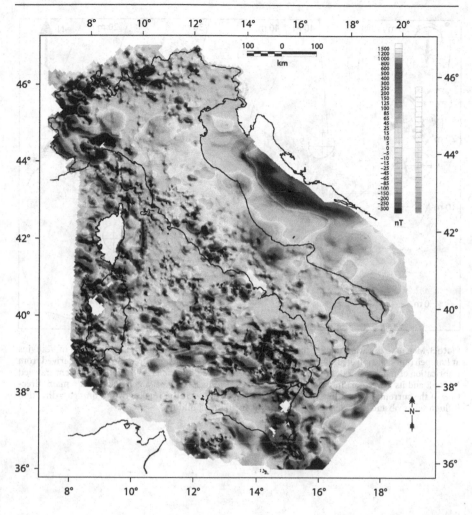

Plate 2. Merged shaded-relief map of the total intensity anomaly of the Earth's field in Italy and surrounding seas derived from ground and shipborne surveys (all data reduced to the sea level) (from Chiappini et al. 2000)

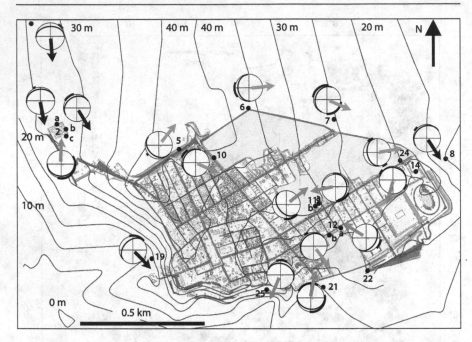

Plate 3. Magnetic lineation in pyroclastic density currents deposits of the A.D. 79 eruption of Vesuvius at Pompeii (after Gurioli et al. 2005). At each sampling site (*dot*) the flow direction (*arrow*) derived from imbrication of magnetic foliation (*great circle in equal-area projection*) is shown. The current traveled N170° E and its lower part interacted with the city buildings. *Color of the arrows: black* = main direction of the current; *green* = direction following the city walls or the road network; *orange* = direction influenced by obstacles or derived from deposits inside the buildings

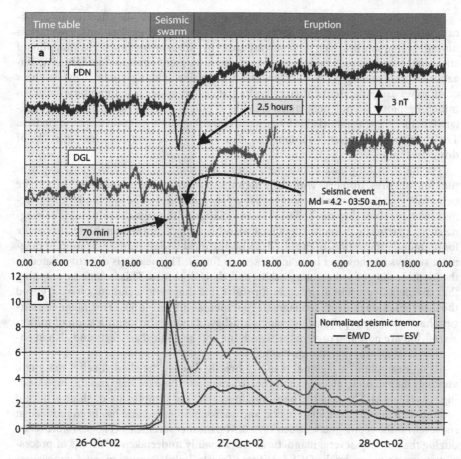

Plate 4. Volcanomagnetic changes at the onset of the October 2002 eruption of Mt. Etna (Italy) (from Del Negro et al. 2004). Plots of 1-minute means of the total field *F* magnitude (**a**) and of seismic tremor (**b**). The first variation in magnetic field at station DGL occurred 70 minutes later than at station PDN (2.5 km to the south of DGL); the recovery phase at DGL started 2.5 hours later than at PDN. A similar delay was observed in seismic tremor at stations EMVD and ESV, located on the southern and northern flanks of the volcanic edifice, respectively

Magnetic anomaly investigation is in general a rapid and cost effective geophysical mapping technique very useful in a variety of geological applications. This task is nowadays frequently undertaken on vast areas by airborne methods. The mapping of structural trends is for example one of the most frequent applications. Magnetic lineations visible in map contours often reflect the strike of elongated intrusive features or at times are indications of large faults concealed under a sedimentary cover. In mineral exploration magnetic maps are able to point to strikes, faults and other similar structures, not always visible by means of a geological inspection from the surface. The shape of the magnetic contour can also be a sign of contacts between rocks of different types.

The most striking case of magnetic applications are related to magnetic exploration for iron ores. Iron ores with their high magnetite/hematite content are easy to be detected by magnetic measurements. Also other minerals can be searched for by magnetic surveys since they can be found in association with magnetic minerals like magnetite or pyrrhotite. Some surveys have also been made in the search for kimberlite for diamond exploration. In the last decades the most widespread application of magnetic surveys has however been in the search for oil. This search is directly related to the identification of sedimentary basins and their thickness. This is done by mapping the basement depths, in many cases it results that magnetic techniques have allowed the determination of depth to magnetic basements with only a 5% error. Magnetic surveys have not only supported petrologic studies, allowing the identification of minerals, oil, but are now also used in the search for archaeological artefacts and also in environmental geophysics studies (Sect. 8.1).

As an example of a regional map, we briefly discuss the shaded-relief magnetic anomaly map of total intensity of the Earth's magnetic field for Italy and the surrounding seas (Plate 2). The map has been obtained after the compilation of shipborne and ground data sets for the Italian region and surrounding seas, and has been reduced at sea level, for the geomagnetic epoch 1979.0. Marine measurements (Fig. 3.9) were taken during the years by several magnetic surveys mainly undertaken by means of precession magnetometers by the OGS Institute (Trieste, Italy). Ground magnetic measurements were taken at more than 2 600 points by teams from Italian universities and scientific institutions. A total of about 50 000 data points finally entered the magnetic data set. Offshore and ground data were reprocessed, reduced to a common geomagnetic reference field epoch, the ground data were reduced to sea level to be consistent with the marine data, and all data were merged to produce the map. The IGRF model was finally used to apply the secular variation correction and to compute the magnetic anomalies. The anomaly field is characterized by a wide range of amplitudes and wavelengths and three major domains can be immediately seen at a first glance. To the north, short-wavelength anomalies line up along the alpine belt, follow the arc of the western Alps and continue southwards along the eastern coasts of Corsica (the white, not surveyed island in the map). They mainly correspond to ophiolites massifs, outcropping in the Alps and at shallow crustal depth below the Tyrrhenian Sea to the East of Corsica. South of latitude 41° N, the anomaly field on the Tyrrhenian Sea is characterized by many anomalies of small extent, related to Pliocene and Pleistocene volcanic edifices, comprising both seamounts and volcanic islands. On the contrary, the

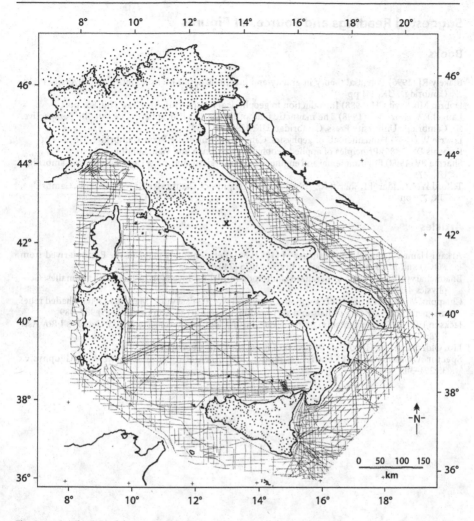

Fig. 3.9. Distribution of the marine profiles and location of the ground magnetic stations for the Italian magnetic survey. The *cross symbol* indicates the location of L'Aquila geomagnetic observatory (lat. 42.38° N, long. 13.32° E) (from Chiappini et al. 2000)

Apennines mountain belt, all along the Italian peninsula, and the Adriatic Sea to the east of it show a long-wavelength pattern. The different magnetic signature is mainly due to the different types of crust: thin and with high heat flow below the Tyrrhenian Sea, thick and with low heat flow below the eastern side of the Italian peninsula and the Adriatic Sea. These features entail an eastward dipping of the Curie isotherm, as also suggested by the difference in the anomalies' background: a generally positive trend characterizes the Adriatic region, whereas in the Tyrrhenian Sea the short-wavelength anomalies stand out from a general negative trend.

Suggested Readings and Sources of Figures

Books

Blakely RJ (1996) Potential theory in gravity and magnetic applications. Cambridge University Press, Cambridge, UK, 441 pp

Dobrin MB, Savit CH (1988) Introduction to geophysical prospecting. Mc Graw Hill, 867 pp

Langel RA, Hinze WJ (1998) The magnetic field of the Earth's lithosphere. The satellite perspective. Cambridge University Press, Cambridge, UK, 429 pp

Lowrie W (1997) Fundamentals of gephysics. Cambridge University Press, Cambridge, UK, 354 pp

Parasnis DS (1986) Principles of applied geophysics. Chapman and Hall, London, 402 pp

Sharma PV (1997) Environmental and engineering geophysics. Cambridge University Press, Cambridge, UK, 475 pp

Telford WM, Geldart LP, Sheriff RE (1990) Applied geophysics. Cambridge University Press, Cambridge, UK, 770 pp

Articles

Arkani-Hamed J, Langel RA, Purucker M (1994) Magnetic anomaly maps of the Earth derived from POGO and Magsat data. J Geophys Res 99:24075–24090

Bhattacharyya BK (1980) A generalized multibody model for inversion of magnetic anomalies. Geophysics 45:255–270

Chiappini M, Meloni A, Boschi E, Faggioni O, Beverini N, Carmisciano C, Marson I (2000) Shaded relief magnetic anomaly map of Italy and surrounding marine areas. Ann Geofis 43(5):983–989

Jackson DD (1972) Interpretation of inaccurate, insufficient, and inconsistent data. Geophys J Roy Astr S 28:97–109

Li Y, Oldenburg DW (1996) 3-D inversion of magnetic data. Geophysics 61:394–408

Spector A, Grant FS (1970) Statistical models for interpreting aeromagnetic data. Geophysics 35:293–302

Chapter 4

Paleomagnetism

Ferromagnetic minerals spontaneously magnetize in the direction of the Earth's field and give the rock a primary remanence J_r. The remanence tends to be preserved over geological times, unless some natural process provides the thermal or magnetic energy needed to modify domain arrangement, and provided no mineralogical transformations occur. Rocks are an archive of the Earth's magnetic history, a source of basic information both for stratigraphy and geodynamics. If the rocks did not undergo any movements after they have been formed, a stratigraphic sequence allows to reconstruct the magnetic field of the past; if instead they did undergo relative movements, the difference between their paleomagnetic directions allows to reconstruct them. The paleomagnetic archive has an enormous advantage over the other geological archives: given the planetary nature of the magnetic field, its data must be consistent at the entire planet's scale, so their time and geographic constraints are very restricted. For example, if rocks in different regions of the globe have the same age, the polarity of their remanence must be the same aside from the petrographic facies, and the directions must concur in the very same point, which corresponds to the magnetic pole for this age.

4.1
Magnetic Remanence in Rocks

The remanence of a rock is called natural remanent magnetization (NRM). It can be acquired through different processes, synchronous to or later than the formation of the rock, which produce a respectively primary or secondary magnetization.

4.1.1
Thermal Remanent Magnetization (TRM)

Thermal remanent magnetization (TRM) is that acquired by igneous rocks while they cool. The TRM model is straightforward: magnetite start crystallizing at temperatures in the order of 1 000 °C and its crystals remain in the paramagnetic state, both in the molten magma and in the solidifying rock. When the temperature of the rock drops below the Curie point (575 °C) the crystals pass first to the super-paramagnetic and then, as cooling continues, to the ferromagnetic state.

Let us examine a simple model: a set of N identical SD crystals shaped as a prolate ellipsoid, oriented parallel to each other and to the Earth's field $F = \mu_0 H_E$. At the temperature T, immediately below the Curie point, the thermal activation energy is high and the energy barriers opposing changes to the orderly state of the domain are low. Magnetization is parallel or antiparallel to the field and changes from one sense to the opposite one in very short times. Parallel magnetization is statistically favored, because it corresponds to the minimum potential energy: the excess ΔN of crystals with parallel magnetization with respect to those with antiparallel magnetization is

$$\Delta N = N \tanh(\mu_0 V J_s(T) H_E / KT) \tag{4.1}$$

where $J_s(T)$ is the magnetization of the crystal at the temperature T, and hence $VJ_s(T)$ is its magnetic moment. The set has a magnetization parallel to H_E, but since its relaxation time τ (Eq. 2.13) is very short, the magnetization of individual crystals varies continuously. As the rock cools down, thermal energy decreases, energy barriers grow higher, τ increases. The transition from parallel to antiparallel becomes progressively more difficult, i.e. less likely, ΔN increases and magnetization becomes ever more stable. At a certain point, the blocking temperature T_b is reached: below this value, the probability that a crystal could change its orientation is extremely low. The situation is frozen and the rock acquires a TRM parallel to H_E, whose intensity is given by the relationship

$$J_{TRM}(T_b) = \Delta N J_s(T_b) = N J_s(T_b) \tanh(\mu_0 V J_s(T_b) H_E / KT_b) \tag{4.2}$$

As temperature continues to decrease, the tanh argument in Eq. 4.2 no longer changes, because the situation is energetically blocked and once ambient temperature is reached

$$J_{TRM}(20\ °C) = N J_s(20\ °C) \tanh(\mu_0 V J_s(T_b) H_E / KT_b) \tag{4.3}$$

This Néel model entails two important consequences, which – as Thellier has demonstrated – agree with experimental data:

1. The TRM acquisition process is reversible, because if the rock is heated above T_b, thermal energy again activates the parallel $\leftrightarrow$ antiparallel transition. Therefore, there is a value of unblocking temperature $T_{ub} = T_b$ where the magnetization acquired during the previous cooling is dispersed.
2. The TRMs of populations of crystals with different T_b, $T_{b1} > T_{b2}$, are mutually independent and they add to each other to give the total TRM of the rock. According to Néel's model, what happens for $T > T_{b1}$ is not recorded by either of the two populations; for $T = T_{b1} > T_{b2}$ the first population acquires its TRM, whilst the parallel $\leftrightarrow$ antiparallel transition is still free in the second population; for $T_{b1} > T > T_{b2}$ the first population remains blocked while the second one continues to be free until $T = T_{b2}$, and is then blocked itself for $T < T_{b2}$. Each crystal population therefore records solely the field present at the instant in which temperature drops below its blocking temperature.

The ferromagnetic crystals of a rock do not constitute a homogeneous population, since mineralogical composition, shape, dimensions, H_c and J_s vary over more or less wide ranges and so does τ. Consequently, T_b varies from grain to grain and the rock is characterized by a spectrum of blocking temperatures. However, Néel's model and Thellier laws remain substantially valid and in good agreement with experimental observations. The curves in Fig. 4.1, derived from the procedure that will be described in Sect. 4.3.2, present three real cases: a rock with very narrow T_b spectrum, close to the value of the Curie point, one with the spectrum split in two clearly separated intervals, the other one with continuous spectrum. As to the direction of magnetization of individual crystals, the statistical approach is somewhat more complex than in the

Fig. 4.1. Blocking temperature (T_b) spectra in igneous rocks; **a** one population of grains with high T_b (rhyolithic ignimbrite, Patagonia); **b** two populations of grains with distinct T_b (diorite, western Alps); **c** heterogeneous population of grains with T_b from ambient temperature to the Curie point (ash-tuff, Aeolian Islands, Tyrrhenian Sea)

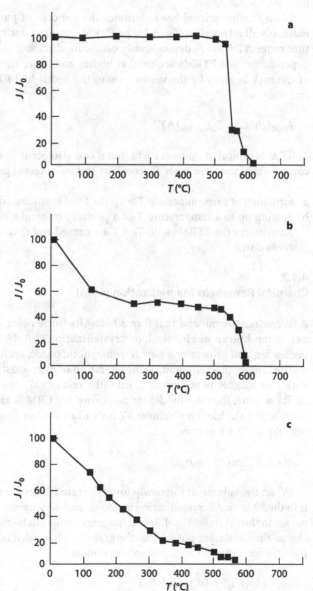

model, where the only possible states where parallel and antiparallel to the Earth's field F. In a rock, the orientation of the crystals and of their directions of easy magnetization is random, except in particular cases. If the rock cools in the absence of a magnetic field, the magnetization of individual grains has a random spatial distribution and TRM is nil; in the presence of F, for each grain the likeliest direction of magnetization is the easy one that is closest to the direction of the field: distribution is no longer random and the resulting TRM is concurrent with F, although it is far less intense than the saturation value, where individual crystals' magnetizations are all mutually parallel.

From Thellier second law originates the concept of partial TRM (PTRM), schematically illustrated in Fig. 4.2. The PTRM acquired during cooling in the temperature range $\Delta T = T_1 - T_2$ depends only on the field present at that moment and it is independent of the PTRMs acquired at higher and lower temperatures. The total TRM of the rock is given by the vector sum of the individual PTRMs acquired in the ΔT_i intervals

$$J_{TRM}(20\ °C) = \Sigma_i J_{i\ PTRM}(\Delta T_i) \tag{4.4}$$

TRM is typical of igneous rocks, but it can also occur in metamorphic rocks under conditions of medium-high T, according to two different processes:

a formation of a new mineral at $T > T_c$: the TRM is acquired during the eventual cooling.
b heating up to a temperature T of a primary mineral without mineralogical transformations: the PTRMs with $T_b < T$ are erased and then acquired again as the rock cools down.

4.1.2
Chemical Remanent Magnetization (CRM)

A ferromagnetic mineral that forms below its Curie point acquires a remanent magnetization known as chemical, or crystallization (CRM). The processes can be the nucleation and growth of a new ferromagnetic phase, such as the formation of a hematite cement in a sediment, or the alteration of a pre-existing phase, such as the oxidation of magnetite to hematite. Since the relaxation time τ of a grain also depends on its volume, the base model for acquiring the CRM is similar to that of the TRM, considering the blocking volume V_b instead of the blocking temperature T_b. Rewriting (Eq. 2.13), we obtain

$$V = 2\ KT\ln(C\tau)\ /\ \mu_0 H_c J_s \tag{4.5}$$

When the volume of the newly formed grain is very small, τ is short and the grain is in the SP state. As crystallization proceeds and dimensions increase, τ grows, the grain passes to the SD state (Fig. 4.3) and magnetization can be considered stable at the geological time scale. Supposing that the grain is spherical, d is its diameter and the other parameters in Eq. 4.5 are constant, we obtain

$$(d_1)^3 / \ln(C\tau_1) = (d_2)^3 / \ln(C\tau_2) \tag{4.6}$$

Fig. 4.2. TRM acquisition during the cooling of an igneous rock. *Gray areas* represent the PTRM acquired in each temperature interval, the *full line* the total TRM, the *arrows* show that the process is reversibile

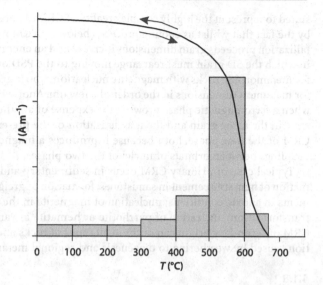

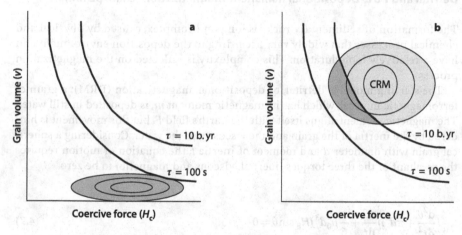

Fig. 4.3. Schematic picture of the CRM acquisition during the growth of ferromagnetic grains. *Heavy lines* represent relaxation time; **a** at beginning of crystallization the grains (*dark gray area*) have a smaller volume than blocking volume and are super-paramagnetic; **b** as the grains grow through blocking volume they change to SD and acquire a stable CRM (*light gray area*) (from Butler 1992)

The growth in τ as volume increases is enormous; assuming $C = 10^9$ s^{-1} (Sect. 2.2.5), a relative increase in the diameter $d_2 / d_1 = 1.4$ entails a growth in τ of 10^{16}. For example, a grain of magnetite is in the SP state as long as $d < 0.03$ μm; when $d = 1$ μm it is in the SD state: supposing that its relaxation time is $\tau = 100$ s, when the grain has grown to $d = 1.4$ μm τ is in the order of 100 Ga.

The model considered for the TRM (all grains with the same characteristics, etc.) applied to the case of the CRM leads to an equation similar to Eq. 4.3. The parallelism between the TRM and CRM acquisition processes is elegant as a theory, but poorly suited to represent the highly complex reality of CRM. A substantial difference is given by the fact that whilst as cooling proceeds below T_b TRM no longer changes, as crystallization proceeds grain dimensions increase and an energy situation can be reached in which the SD grain must rearrange moving to the PSD or MD state. This is typical for metamorphic rocks with magnetite nucleation, whose grains maintain the SD state for maximum dimensions in the order of a few μm. Another complex situation occurs when a ferromagnetic phase grows at the expense of another one: the two phases coexist in the same grain and the magnetization of the pre-existing one influences the CRM of the new phase, both because it produces a magnetic field and for possible couplings between orbitals of nuclei of the two phases.

Typical case of primary CRM occur in sedimentary and metamorphic rocks: formation of hematite cement in sandstones; formation of greigite in marine or lake sediments in anoxic conditions; nucleation of magnetite in the peridotite → serpentinite transformation; nucleation of pyrrhotite or hematite in various metamorphic facies. CRM of secondary origin can occur in all types of rocks and at all scales, from oxidation caused by weathering to diagenesis and regional metamorphism.

4.1.3
Detrital and Post-Depositional Remanent Magnetization (DRM, pDRM)

The formation of sedimentary rocks is long and complex, caused by physical and chemical processes that widely vary according to the deposition environment and have a relatively long duration. This complexity is reflected on the magnetization process.

The starting model is detrital, or depositional, magnetization (DRM): a grain of ferromagnetic mineral, which has its magnetic moment m, is deposited in still water. The magnetic moment aligns itself with the Earth's field F, but this movement is hindered by the inertia of the grain and the viscosity of the water. Considering a spherical grain with diameter d and moment of inertia I, the equation of motion requires the resultant of the three torques (inertial, viscous and magnetic) to be zero

$$I\frac{d^2\theta}{dt^2} + \pi d^3 \eta \frac{d\theta}{dt} + \frac{\pi}{6}\mu_0 d^3 J H_E \sin\theta = 0 \tag{4.7}$$

where θ is the angle between the two vectors m and $F = \mu_0 H_E$, η is water viscosity, J the magnetic moment per unit volume, i.e. the remanent magnetization the grain acquired when the mineral formed in the parent rock. Figure 4.4 shows the relative importance of the three moments in the case of spheroidal grains. Introducing in Eq. 4.7 numerical values that are representative of real cases, one derives that the alignment of m with F is achieved in times in the order of a few seconds. However, the equilibrium may be disturbed as the grain reaches the bottom and a mechanical torque tends to position it parallel to the bedding plane. The final situation depends on grain size (Fig. 4.4): with dimensions <10 μm, the magnetic torque prevails and alignment with the Earth's field is achieved, with dimensions >0.1 mm the mechanical torque prevails and the directions of individual grains are dispersed randomly. The direction of the Earth's

Fig. 4.4. Torques exerted on spheroidal magnetite grains settling in still water (from Dunlop and Özdemir 1997)

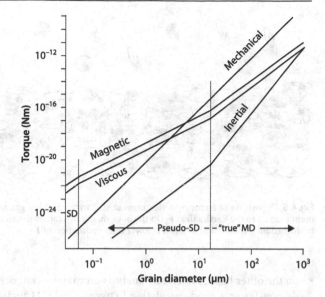

field is thus recorded by the DRM in claystones and fine siltstones, while magnetization becomes progressively more chaotic in coarse siltstones and sandstones.

It should be stressed that DRM is a passive orientation under the action of an external field on already magnetized grains: the paleomagnetic study of sedimentary rocks, therefore, had a fundamental role in demonstrating the existence of polarity reversals, since a DRM with reverse polarity can only have been acquired in the presence of a reverse polarity field.

The reality of DRM is not as idyllic as in the model discussed above: if the water is not still, the current tends to orient the grains; if the bottom is not horizontal, they can roll over; but above all, the spheroidal grain assumption is not realistic. Grains are usually elongated or flattened and are deposited on the bottom with their greatest dimension parallel to the bedding (Fig. 4.5). The orientation of the magnetic moments is thus offset by an effect that is random for declination, but systematic for inclination, which is smaller than that of the Earth's field. This effect, called inclination error or shallowing, has been observed in many cases and has generated numerous studies on sediments, both natural and re-deposited in the laboratory under controlled conditions, to understand how much confidence can be given to the paleomagnetic data obtained from sedimentary rocks. The results are interesting:

- on the one hand, the inclination error is observed both in natural and re-deposited sediments (Fig. 4.6) and can be described by the relationship

$$\tan I_s = f \tan I_H \tag{4.8}$$

where I_H is the inclination of the external field (terrestrial or created in the laboratory) and I_s the inclination of the DRM of the sediment, while f is an experimental coefficient, in many cases ≈ 0.4 to 0.6.

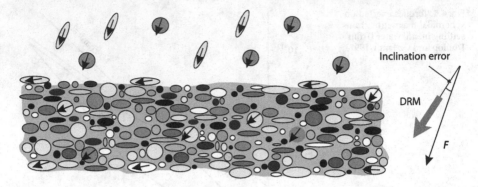

Fig. 4.5. Deposition of ferromagnetic grains in still water. Settling grains arrange their magnetic moment parallel to the Earth's field *F*. On the bottom, gravity causes the elongated grains arrange parallel to the bedding. Inclination of DRM is lower than inclination of *F*

- on the other hand, systematic analysis on many recent oceanic sediments (Fig. 4.7) demonstrates a good correlation between the DRM inclination and that expected according to the GAD model at the latitude λ^1 of the sampling site, $\tan I_s \approx 2 \tan \lambda$.

From these observations, the model of post-depositional magnetization (pDRM) was devised; it starts from the fact that newly sedimented grains form a mud layer with a high water content, which is progressively eliminated as the weight of overlying sediments increases. The grains are still free to move for a certain period of time, which depends on porosity, which in turn depends on the grains of silicate or carbonate minerals that constitute the must abundant fraction of the sediment. For equal weights, the dimensions of these grains are greater than those of ferromagnetic grains, which are still free to move inside water saturated pores and hence to reorient because of the torque exerted by the Earth's field (Fig. 4.8). The finer the grains, the longer they retain their freedom of movement. For dimensions below a few µm, the Brownian motion must be taken into account: it is due to the energy transferred by water molecules to ferromagnetic grains. The situation is similar to that of paramagnetic substances: Brownian motion scatters magnetic moments, but they statistically tend to align themselves in the direction of the magnetic field. The phenomenon can be described by an equation similar to Eq. 2.3

$$J = J_s L(\alpha) \approx J_s (\mu_0 V J_r / 3\ KT) H_E \tag{4.9}$$

---

[1] The geocentric axial dipole (GAD) model corresponds to the SHA term $n = 1$, $m = 0$. It assumes that the Earth's magnetic field is that of a dipole whose axis coincides with the rotation axis of the planet. The study of SV shows (Chap. 1) that the Earth's field averaged over times greater than a few hundreds kyr may be assumed as a GAD field. The symbols for latitude and longitude (φ, λ) used in Chap. 1 follow the convention of most books on geomagnetism; those (λ, φ) used in this chapter follow the convention of most books on paleomagnetism. We used the two conventions in order to make easier the comparison of equations between different books on the same subject.

Fig. 4.6. Relationships between inclination of remanence (I_s) in re-deposited sediments and inclination of the laboratory controlled applied field (I_H). Glacial sediments (*1, dot*) conform to the DRM model of inclination shallowing ($I_s < I_H$), deep-sea sediments (*2, square*) to the pDRM model ($I_s \approx I_H$). (modified after Verosub 1977 and Irving and Major 1964)

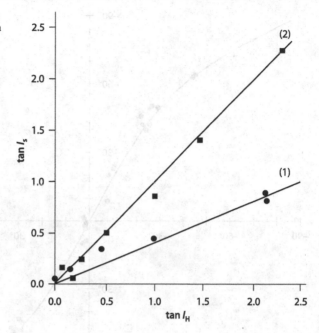

where J is observed magnetization, J_s saturation magnetization, i.e. with all the grains' moments mutually parallel, L the Langevin function, V the volume of the grains and J_r their remanent magnetization.

When porosity and water content are further reduced, grain movements are no longer possible: this process requires a certain time, known as lock-in time. Lock-in is not the final point, because the first phases of diagenesis are still left. As overburden increases, compaction tends to produce a planar fabric and grains tend to resettle according to the bedding. If the rock still has a certain permeability, fluid circulation may vary the chemical conditions, especially of oxy-reduction. The original detrital ferromagnetic minerals can be destroyed and new ferromagnetic minerals may be formed. The resulting CRM is nearly synchronous to the DRM.

4.1.4
Isothermal Remanent Magnetization (IRM)

A ferromagnetic substance subject to the action of an external magnetic field retains a remanent magnetization even when the field is removed, as shown by the hysteresis cycle (Sect. 2.2.2). This isothermal magnetization (IRM) is easily produced in the laboratory and it is used to characterize the ferromagnetic minerals of the rocks (see Sect. 4.3.3). The IRM of natural origin is due to lightning strikes: an electrical current with intensity in the order of ≈ 100 A flows in the ground for some milliseconds and generates a very strong magnetic field, which imparts to the rock an IRM with random direction and very high intensity, which can approach the saturation value J_{rs}. An outcrop struck

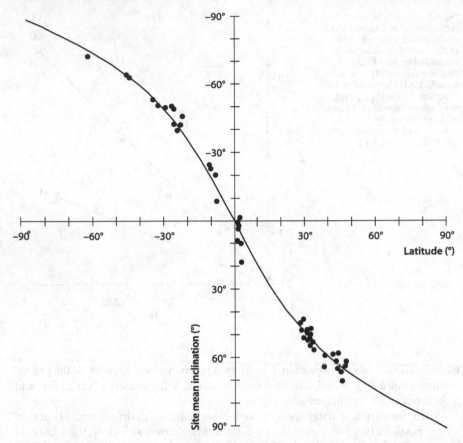

Fig. 4.7. Inclination of remanence from deep-sea cores vs. coring site latitude. Symbols: *dot* = remanence inclination; *full line* = inclination expected according to the GAD model (from Opdyke 1972)

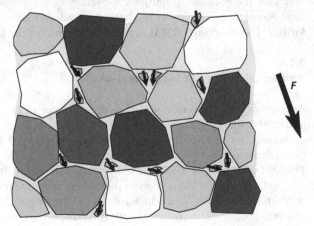

Fig. 4.8. Shematic picture of pDRM acquisition. Small ferromagnetic grains are free to move within the pore spaces of a sediment and tend to reorient in the direction of the field *F*

by lightning can thus cause strong compass deflections and it is no surprise that this phenomenon had been observed in the nineteenth century, when the term *fulgurite* (from the latin *fulgur* for lightning) was coined to indicate rocks that had been struck by lightning. Therefore, it is advisable to avoid sampling lightning-prone sites, as topographic highs.

4.1.5
Viscous Remanent Magnetization (VRM)

Remanent magnetization corresponds to an orderly state that lasts over time, because it is bounded by energy barriers. However, there is always a certain probability that barriers are overcome, the magnetic domains change their arrangement and the magnetization gradually changes over time. In the case of rocks, the new magnetization tends to be oriented in the direction of the Earth's field. Rocks can thus acquire a viscous magnetization (VRM), according to a law of the kind

$$J_{VRM} = S \log t \tag{4.10}$$

where S is the coefficient of viscosity, which depends on the characteristics of the domains, and t is the time during which the rock was subjected to the field. It is clear from Eq. 4.10 that the present-day field has a major effect. VRM is a secondary magnetization and it is nearly always to be considered as noise, which can be eliminated with appropriate procedures (see Sect. 4.3.2). However, when S is high, the VRM acquired in a few seconds can be large enough to cause a change in the value of remanence during the time required to perform its measurements.

4.1.6
Other Remanent Magnetizations

The remanent magnetizations discussed in the previous sections are the most important ones in geological applications; however, many other physical processes, natural and artificial, can influence the orderly state of the magnetic domains and thus produce a remanence. Let us briefly mention:

a Piezoremanent magnetization (PRM). The application of a stress entails mechanical deformations of the grains, which are transferred to the walls of the domains, causing non-reversible changes and hence a remanence.
b Drilling induced magnetization (DIRM). Oftentimes samples obtained from oceanic or continental cores have a spurious magnetization that can be correlated to the magnetic field of the steel core barrel and the mechanical stress caused by drilling.
c Anhysteretic magnetization (ARM). A sample that is simultaneously subject to an alternating magnetic field that decreases from a maximum value to zero and to a steady field, acquires a remanence parallel to the steady field. ARM measurements are a routine laboratory technique used to characterize ferromagnetic minerals (see Sect. 8.2).

4.1.7
Timing of Remanence Acquisition

The aim of most paleomagnetic studies is to identify the primary magnetization, synchronous with the formation of the rock. Therefore, the meaning of synchrony should be discussed a bit more in depth. Figure 4.9 outlines the time interval in which primary magnetization is acquired in various types of rocks. In the case of sub-aerial volcanic rocks, things are simple: cooling takes place in a time interval of a few hours to a few years and from the geological viewpoint the TRM has the same age as the rock. The isotopic age of the rock, determined for example with the ^{40}Ar/^{39}Ar method, is transferred unaltered to the TRM. If we consider intrusive rocks, cooling requires longer times and the isotopic age is no longer that of emplacement, but rather the one in which the isotopic system used for dating was closed. Similarly, the age of the primary TRM is the one in which the temperature dropped below the blocking temperature. For example, the Rb/Sr system closes in biotite around 350 °C; if the TRM of the rock is carried by magnetite grains with T_b exceeding 500 °C, its age is certainly older than the Rb/Sr age. Intrusion, closing of the isotopic system, blocking temperature thus represent three distinct episodes of the history of the rock, which can be offset even by millions of years when the cooling process is very slow, as for intrusions in the lower crust. The case of metamorphic rocks is similar; crystallization or cooling processes occur inside the Earth's crust and continue over long time intervals. The lock-in time of sedimentary rocks depends primarily on the sedimentation rate, which varies over many orders of magnitude, from a few cm/yr in large lakes of tectonic origin to a few mm/1 000 yr in deep oceanic plateaux. A standard specimen with dimensions of 2–3 cm thus corresponds, in terms of sediment accumulation and subsequent compaction, to a time interval ranging from a few tens to a few thousands of years, with a very different chronological significance.

Hence, the chronological relationship between the formation of the rock and the acquisition of its remanence must be carefully evaluated. If the two processes are rapid,

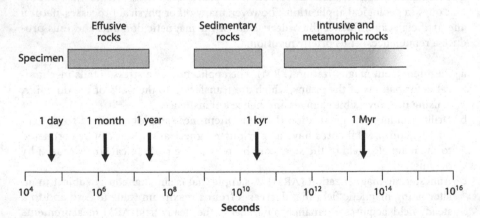

Fig. 4.9. Schematic outline of the remanence acquisition time in various types of rocks (time scale in seconds). *Gray boxes* refer to the time interval recorded in a specimen

the situation is simple: the rock and its primary magnetization have the same age. When they are diluted over time, things are not immediate and they become progressively more complicated as time becomes longer, also because the direction of the Earth's field can undergo changes, such as a polarity reversal, which are recorded in the same specimen. Rapidity must be evaluated also as a function of the age of the rock and hence of the chronological resolution to be obtained. A lock-in time of a few ten thousands of years is long for Pleistocene sediments, short for Palaeozoic sediments.

4.2
Sampling Techniques

Paleomagnetic samples can be taken with hammer and chisel, as geologists do, or by directly coring on the outcrop, obtaining what are respectively called hand samples and cores. Both techniques have their supporters: if applied correctly, they are wholly equivalent and the choice depends on many factors, not all of them under the paleomagnetist's control, such as:

- consistency of the rock. Loose or poorly consolidated rocks, such as some tuffs or recent sediments, are difficult to core. Large hand samples ($10 \times 10 \times 10$ cm) can be taken from the outcrop, consolidated in the laboratory and then cut to specimens. Another technique consists of gently pressing on the rock a cubic box made of plastic (diamagnetic).
- availability of water. During coring, water is essential to cool the diamond bit and clear away the scrap.
- environmental value. In natural parks, sites of special geological interest, wild regions such as Antarctica the holes left by coring are not very compatible with the environment to be preserved.

To orient a hand sample, the strike and dip of a face are measured exactly as with bedding; for cores, a special tool is used, formed by an orientation table hinged on a slotted tube: the tube is inserted in the hole making the slot to coincide with the vertical plane through the axis of the core. Orientation is done with compass and clinometer. Since the reference system is geographic north, it is essential to know the magnetic declination at the sampling site either by deriving it from maps or better by measuring it with the Sun compass. In the case of intensely magnetized rocks, Sun orientation must be measured for each individual sample; in the case of volcanoes, for example, it is not uncommon to observe erratic declination changes in the order of ±5° between points in a same outcrop.

Sea and lake bottom sediments are sampled as piston cores with large diameter, $\varnothing \approx 10$ cm. These cores are often not azimuthally oriented and only the up/down side of the specimen is known. Therefore, paleomagnetic measurements will only provide remanence inclination, provided the core penetrated vertically.

Samples are then cut in the laboratory to prepare the specimens to be measured; the widespread use of commercial instruments has led to the nearly complete standardization in the shape. Specimens are either cylinders with diameter $\varnothing = 25.4$ and height $h = 22$ mm or cubes with side of 20–25 mm.

The plan of the sampling is the most delicate point of any paleomagnetic study, from which the quality and "weight" of the results depend. It must be founded on a thorough knowledge of geological and petrographic literature and a good familiarity with cropping out conditions, and it must be best suited to the problem in question. For example, the results of a tectonic study heavily depends on the areal distribution of the sampling sites. Fresh rock outcrops should always be preferred, because weathering easily causes secondary magnetizations due to oxidation and formation of Fe hydroxides. The best outcrops for paleomagnetism are natural cuts, although oftentimes one has to make do with road cuts. Quarries can be deceiving: the rock is fresh, but use of explosives may have caused PRM.

4.3
Laboratory Techniques

The goal of paleomagnetic measurements is to measure remanent magnetization, identify its various components and find out the ferromagnetic minerals carrying each individual component. There are very many instruments and techniques, and we will limit ourselves to basic aspects.

4.3.1
Remanence Measurements

Remanent magnetization J_r is a vector: the measurement must therefore provide magnitude (J_r) and direction, defined with the same declination and inclination angles (D, I) used for the Earth's field. The three components of J_r are obtained in the specimen reference system (x, y, z) and then transformed to the geographic system using the field orientation of the sample. Magnetometers do not measure J_r directly, but the magnetic field B the specimen generates in the surrounding space. Assuming that the field is that of a dipole, the magnetic moment M is derived and from it, assumed to be homogeneous within the specimen, J_r is computed. To satisfy the first hypothesis, the shape of the specimens must approximate the sphere as closely as possible: standard shapes are the cube and the cylinder with height/diameter ratio $h / \varnothing \approx 0.9$. The lack of homogeneity of M, and hence of J_r, in the rock is often limited and the related errors can be reduced by repeating the measurements in different specimen/sensor relative positions. Sensor and specimen are appropriately shielded against external magnetic fields, to prevent the presence of induced magnetization J_i (Earth's field) and minimize noise (artificial fields).

The spinner magnetometer works based on the principle that a revolving magnetized specimen generates an alternating e.m.f. in a coil. The signal frequency is equal to revolution rate, its amplitude is proportional to the component of the magnetic moment perpendicular to the axis of revolution. The spinning system generates a reference signal, whose phase allows to split the signal in two orthogonal components, proportional to two components of the magnetic moment. Repeating the measurements with the specimen in different, mutually orthogonal positions, the three components are obtained. The cryogenic magnetometer is based on superconductivity effects. Its sensor is called SQUID (Superconducting Quantum Interference Device). A ring of superconductor material is maintained below the critical temperature. When a magnetized specimen is moved close, the magnetic flux Φ linked to the ring changes, and the change $\Delta\Phi$ induces in the ring an electrical current, whose intensity depends on the component of the field B of the specimen parallel to the axis of the ring. The sensors are installed inside a Dewar vessel, shaped in such a way as to enable the specimen kept at ambient temperature to be moved close to it. Having three mutually orthogonal SQUIDs, the three components of B can be measured all in one.

To reduce measurement errors and any errors due to the non homogeneity of J_r, each component must be measured at least once in the two opposite directions ($\pm J_x$, $\pm J_y$, $\pm J_z$). Precision can be reduced by viscous components with relaxation times in the order of a few seconds.

Modern instruments provide reliable measurements of magnetic moments $M \approx 10^{-10}$ A m^2, and hence specimens with magnetization $J_r \approx 10^{-5}$ A m^{-1}, as for example many carbonate rocks. Basic effusive rocks can exhibit values up to $J_r \approx 10$ A m^{-1}; higher values are observed in particular cases (mineralizations, lightning strikes, ...).

4.3.2
Demagnetization

Rocks acquire their natural magnetization (NRM) through various processes, which act at the time they form (primary magnetization) as well as later (secondary magnetizations). The NRM can thus consist of more than one component, each with a different geological and chronological meaning. The measured vector (J_{NRM}) is the resultant of the various components, so the problem of identifying them needs to be confronted. This is mathematically impossible, because a vector can be resolved into components in infinite different ways. Resolution can instead be possible through physics, based on a simple principle. If a rock has multiple magnetization components, each of them involves a different population of ferromagnetic grains: if they have different magnetic properties, for example different T_b or H_c, the components can be cancelled out one at a time, and this allows to isolate them (Fig. 4.10).

To cancel a magnetization, thermal or magnectic energy must be provided. Thermal demagnetization is based on the reversibility of PTRM (Sect. 4.1.1). It is performed with an oven shielded from all external magnetic fields and built in such a way that the magnetic fields caused by the heating current cancel each other out. After measuring J_{NRM}, the specimen is heated to the temperature T_1. The magnetic domains with $T_b \leq T_1$ lose their magnetization, which they reacquire when the specimen cools down to ambient temperature, always within the magnetic shield. Since there is no field, there is no preferential direction, so each domain is magnetized in its easy direction and the resultant of the domains with $T_b \leq T_1$ is zero. The measured remanence, then, is $J(T_1) \leq J_{NRM}$. Heating to temperature T_2, the magnetization of the domains $T_b \leq T_2$ is cancelled and after cooling $J(T_2) \leq J(T_1)$ is measured. The process is repeated step by step until $J(T_n) = 0$, i.e. the maximum value of T_b has been reached. This value can coincide with the Curie point but also be lower. The curves of Fig. 4.1 were obtained in this way.

Demagnetization in alternating fields (Af) is wholly similar, and is based on the magnetic hysteresis principle. Instead of the oven, a solenoid shielded from external magnetic fields is used. It carries an alternating current, which generates an alternating field that is parallel to the axis of the solenoid and has a peak value of H_1. Domains

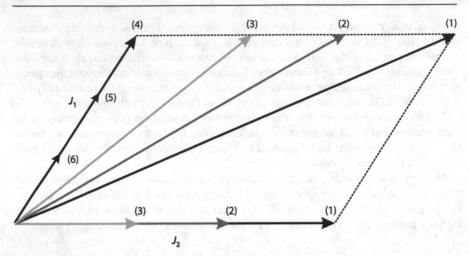

Fig. 4.10. Stepwise demagnetization of a NRM consisting of two components with different blocking temperature or coercivity spectra. As the less stable component J_2 is erased, the direction of the measured remanence varies (*Steps 1 to 4*); when J_2 is completely erased, only the more stable component J_1 survives; direction does not change any more, intensity progressively decreases (*Steps 4 to 6*)

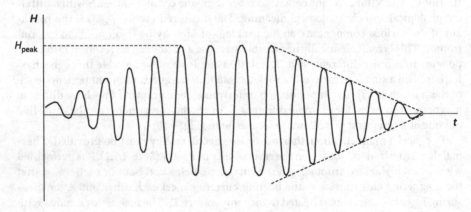

Fig. 4.11. Alternating field (Af) demagnetization. The intensity of the field increases up to and for a few seconds remains at a maximum value, H_{peak}; then linearly decays to zero

with $H_c \leq H_1$ follow the oscillations of the field; when it is made to decrease (Fig. 4.11), each domain is remagnetized as the field passes at its H_c value. The two opposite senses of the alternating field are two preferential directions, and when $H = 0$ domain magnetization is statistically distributed half in one sense, half in the opposite, thereby yielding a zero resultant. A second step is then done at peak field H_2 and so on as in the thermal procedure. Figure 4.12 is the equivalent of Fig. 4.1 and it shows some real cases of rocks with different coercivity spectrum. Af demagnetization is routinely performed to maximum values of a few hundreds of mT and it does not always cancel out the whole J_{NRM}. A widely used parameter to characterize the coercivity of a rock is

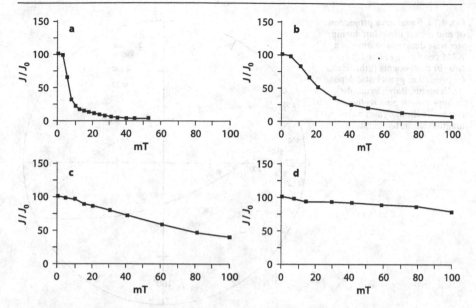

Fig. 4.12. Af demagnetization in rocks with different coercivity spectra; **a** serpentinite with mainly MD magnetite (Lanzo Massif, western Alps); **b** dolerite sill with mainly PSD magnetite (Victoria Land, Antarctica); **c** hematite-rich clayey sediment (Altai Mountains, southern Siberia). Curve **d** refers to a red-coulored film carrying hematite as pigment (mural painting in Vatican Palaces, Rome)

the median destructive field (MDF), which represents the peak value of the alternating field that erases 50% of the J_{NRM}.

The curves of Figs. 4.1 and 4.12 inform us on how remanence intensity varies during demagnetization, but yield no information on the direction. In paleomagnetism the equal-area, or Schmidt, stereographic projection is used to represent the directions. Figure 4.13 is readily interpreted. If the NRM has a single component, the direction of J_r measured after every demagnetization step does not change and all the corresponding points in the net fall close to each other. If instead the NRM consists of two components, the direction of J_r changes as one is cancelled: the points fall along a great circle, because two vector components define a plane and as one is cancelled their resultant moves within this plane. However, a more thorough knowledge of what takes place during demagnetization requires the ability to evaluate simultaneously changes in intensity and direction and this is done using the orthogonal projection diagrams devised by Zijderveld. Let us examine them in some detail, because they are the key that in many cases allows to understand how the NRM is constituted and to isolate its individual components.

Figure 4.14 shows the construction of a Zijderveld diagram:

a The J_r measured after each step (1, 2, ...) is drawn as a vector from the origin of a Cartesian system.

b The vector is projected orthogonally onto the horizontal plane N/S-E/W: the full dot corresponds to the head of the vector J_r. The projection represents the horizontal component and hence its angle with the axis N/S yields the declination D.

Fig. 4.13. Equal-area projection of remanence direction during stepwise demagnetization of a NRM consisting of one (*a*) or two (*b*) components (lithic clasts embedded in pyroclastic deposits, Vesuvius, Italy). Symbols: *full/open dot* = positive/negative inclination, *figures* = demagnetization temperature

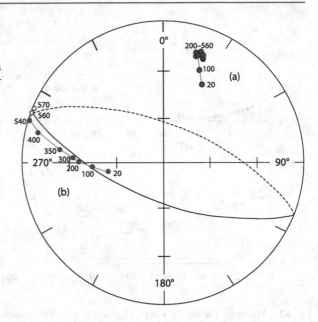

c The vector is projected orthogonally onto a vertical plane (E/W-Up/Down in the example): the open dot corresponds to the head of the vector J_r. The projection gives a component, which forms an angle I^* with the horizontal plane. I^* is called apparent inclination and its value depends on the angle between the projection plane and J_r. When the vertical plane is oriented E/W, the actual inclination is given by $\tan I = \tan I^* |\sin D|$.

d The two planes are superposed and the dots corresponding to the head of the vector J_r are drawn. The curve that joins the full dots shows how D changes as intensity decreases, the one that joins the open dots shows how I^* changes. The two curves point towards the origin of the diagram, which corresponds to the value $J_r = 0$.

Figure 4.15 shows some real cases. If the diagram consists of two lines which point towards the origin, the two values D, I are constant and hence the NRM consists of a single component. If there are two components and their spectra of T_b or H_c are separate, the demagnetization first cancels the less stable component (depending on the type of demagnetization called low-T_b or soft), after the more stable one (high-T_b or hard). The values D, I vary as long as the soft/low-T_b component is cancelled, and are then constant while the hard/high-T_b is cancelled. Each of the two curves of the diagram is a broken line: the first segment corresponds to the less stable component, the segment that tends to the origin corresponds to the more stable one. If the spectra overlap partially, the broken lines assume a curvilinear trend and the stable component can still be recognizable from the profile of the points corresponding to the higher demagnetization steps.

In most cases, the purpose of the demagnetization is to isolate the more stable magnetization component, i.e. the one that corresponds to the higher values of T_b or H_c, which is called characteristic (ChRM). Hence the expression of *magnetic cleaning*,

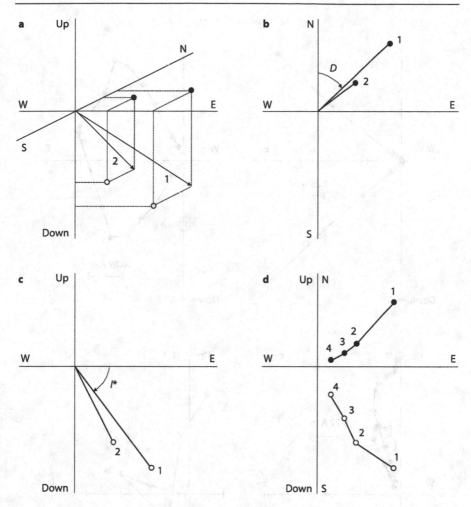

Fig. 4.14. Construction of Zijderveld diagram (see text for further explanation). Symbols: *full dot* = projection of the arrow's head onto the horizontal plane; *open dot* = projection of the arrow's head onto the E-W vertical plane; D = declination; I^* = apparent inclination; *figures* = demagnetization steps

is often used to indicate the demagnetization process. Components with lesser stability are of interest in many applications, whilst those cancelled in the very first few steps often correspond to VRM. The direction (D, I) of each component is calculated interpolating the points of the curve that correspond to its stability interval.

If the superposition between the spectra of T_b or H_c of two components is extensive, the direction of their resultant continually varies during demagnetization and they cannot be separated. Some information can be obtained with other interpretation techniques, such as remagnetization circles. We have seen that the vectors corresponding to two components define a plane, which in equal-area projection is represented by a great circle along which their resultant moves (Fig. 4.13). The simplest case

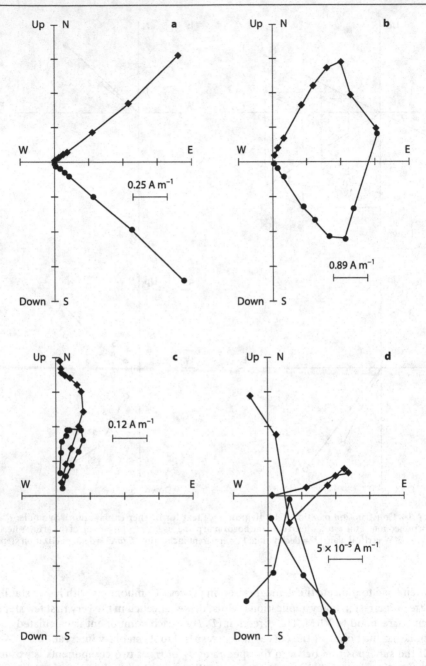

Fig. 4.15. Examples of Zijderveld diagrams. Symbols: *square* = declination; *dot* = apparent inclination; *scale bar* = remanence intensity; **a** single component remanence (basic granulite, western Alps); **b** two components with distinct spectra (diorite, western Alps); **c** two components with overlapping spectra (lithic clast embedded in pyroclastic deposits, Vesuvius, Italy); **d** three components (Cretaceous limestone, *Scaglia bianca*, central Apennines)

is the one in which the NRM is formed by a primary component, equal in all specimens of a site, and a secondary component, variable from specimen to specimen. The
orientation of the great circle therefore changes from specimen to specimen (Fig. 4.16),
but the circles intersect along a common direction, i.e. that of the primary component
present in all specimens. The ChRM direction is thus not provided by the individual
specimens, but from the set of the specimens of the site.

4.3.3
Identification of Magnetic Minerals

In order fully to understand the geological meaning of the ChRM and secondary magnetizations, it is essential to know which minerals carry them and which are their main
magnetic properties. Classic mineralogical techniques, such as X-ray diffraction, reflected light microscopy, micro-probe analysis are made difficult by the very low content of ferromagnetic minerals and by their very small dimensions, even <1 μm. They
can be helpful in the case of igneous rocks, but they are seldom useful for sedimentary rocks. Methods based on the diversity of magnetic properties have greater discriminating power and, in addition to identifying the mineral, they can also characterize its magnetic state (SD, PSD, MD) and useful parameters to evaluate the origin
and the stability of the remanence (coercivity, Curie temperature, etc.). These methodologies are based on the observation of changes of magnetic properties when a
sample is subjected to changes in temperature, applied magnetic field or both at the
same time. They are very numerous and often based on phenomena that require much
more advanced knowledge of ferromagnetism than the basic notions we reviewed in

Fig. 4.16. Remagnetization circles. Each great circle is defined
by the two remanence components of an individual specimen; the circles of all the specimens from a site intersect along
a common direction, given by
the *black dot* with ellipse of
confidence

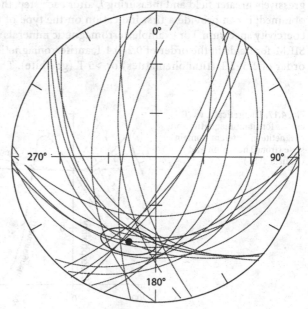

Chap. 2. Therefore, we will just provide a few examples pertaining to some of the most commonly used techniques.

Temperature analyses are performed from the Curie point down to a few degrees Kelvin. Temperature acts both directly on the exchange forces and on the characteristics of the crystal, modifying the lattice and thus causing a change in the magnetic properties. Discontinuous transitions from one state to another occur for example at about –15 °C in hematite (Morin transition) and around –150 °C in magnetite (Verwey transition). The determination of the Curie point T_c is based on the fact that for $T > T_c$ ferromagnetic minerals lose their characteristic properties and behave as paramagnetic: remanent magnetization J_r disappears and the magnetic susceptibility κ collapses to very low values. The sample is inserted in a miniature oven and the whole is subjected to the action of a magnetic field that generates the signal to be measured. In susceptibility measurements, the oven is positioned inside the coil used to measure κ. The value of the Curie point depend on the mineral, whilst the shape of the curve of κ vs. T also depends on the type of domain (Fig. 4.17). To measure the saturation magnetization J_s, the oven is positioned within a non uniform magnetic field, which magnetizes the sample and tends to displace it. The intensity of J_s decreases with increasing temperature and is measured monitoring the feedback force needed to keep the sample in the initial position. Abrupt changes in the slope of the curve of J_s versus T often point to the occurrence of distinct ferromagnetic phases with different Curie point.

The application of an external field provides indications on magnetization and coercivity: the measurements can be performed either in the presence of the field (hysteresis curves) or after cancelling it (IRM measurements). The IRM can advantageously be imparted with a pulse magnet, in which a continuous current flows for a few milliseconds inside a solenoid. In this way, high intensity fields can be produced without any particular technical problems. Subjecting a sample to a series of pulses with progressively greater field and measuring J_r after each step, the IRM acquisition curve is obtained; it can provide a first indication on the type of the minerals and on their coercivity spectrum. For example, ferromagnetic minerals reach the saturation value, SIRM, for fields in the order of 0.2–0.4 T, antiferromagnetic minerals for fields in the order of 1.5–2 T (titanohematites) or >5 T (goethite). The coexistence of different

Fig. 4.17. Theoretical κ vs. T curves for titanomagnetite and magnetite in different domain state (from Thompson and Oldfield 1986)

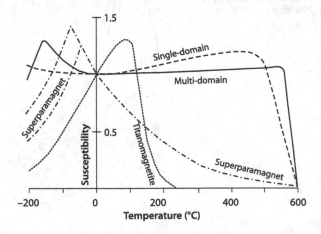

phases can be highlighted with the procedure proposed by Lowrie. The sample is subsequently magnetized at three decreasing field values in three orthogonal directions, $H_x > H_y > H_z$. The three components of the IRM are thus carried by grains with high (J_x), intermediate (J_y) and low-coercivity (J_z). A subsequent thermal demagnetization gradually removes the three components and provides indications on their blocking temperature spectrum and Curie point (Fig. 4.18).

Hysteresis curves provide a greater number of parameters (Sect. 2.2.2) and therefore are richer in information than IRM curves. A widely used instrument is the vibrating sample magnetometer. A sample is made to vibrate in the presence of a uniform magnetic field H produced by an electromagnet: the flux linked to one or more pickup coils changes and induces in the coils an e.m.f., which depends on the moment M of the sample, which in turn is a function of the intensity of H. In addition to mineralogical indications (Fig. 4.19), it is also possible to identify the various types of magnetic domains, for example using Day plots, which show the J_{rs}/J_s ratio as a function of the H_{cr}/H_c ratio.

Fig. 4.18. Thermal demagnetization of three-component IRM. Symbols: *triangle* = low-coercivity component; *dot* = intermediate-coercivity component; *square* = high-coercivity component; **a** andesitic lava (Aeolian Islands); **b** rhyolithic ignimbrite (Patagonia); **c** Eocene clayey sediments (eastern Kazakhstan)

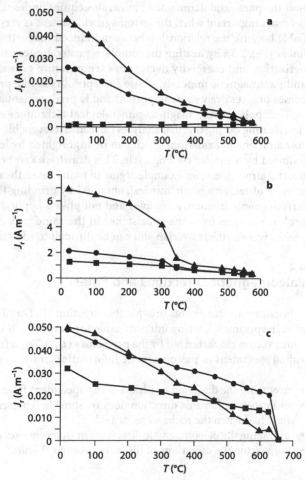

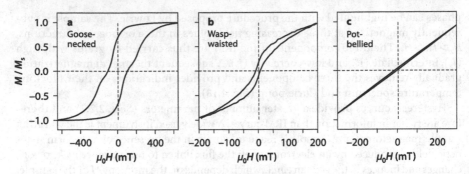

Fig. 4.19. Hysteresis loops of mixtures of different magnetic minerals (**a**: magnetite + hematite) and different domain states (**b,c**: SD + SP magnetite) (from Tauxe 1998)

In the case of a rock, the hysteresis cycle is also influenced by the induced magnetization the para- and diamagnetic minerals acquire under the action of H. Their effect becomes important when the ferromagnetic content is very low, but it is easily identifiable because the relationship between induced magnetization and field intensity is linear (Fig. 2.3). By heating the sample, repeated hysteresis cycles indicate how magnetization and coercivity decrease as temperature grows; any contribution of dia- and paramagnetic minerals, whose susceptibility is independent of temperature, becomes progressively more important and is predominant above the Curie point.

The application of magneto-mineralogical techniques encounters several difficulties. Heating during thermal analyses can cause irreversible transformations. The most common one is oxidation, which can be highlighted by letting the heating cycle be followed by a similar cooling cycle. This drawback can be limited by operating in a neutral atmosphere, for example argon. In many cases, the measurements need a high content of ferromagnetic mineral, obtained by grinding the rock and separating its ferromagnetic fraction, a complex and not always applicable operation. The greatest problem derives from the coexistence in the same rock of different ferromagnetic phases whose effects overlap and can be difficult to identify and separate.

4.4
Paleomagnetic Directions and Poles

Laboratory measurements provide the direction (D, I) and intensity (J_r) of the ChRM of each specimen. Leaving intensity aside for now, to each specimen can be associated a unit vector, characterized by the two values (D, I). To go from the vectors of the individual specimens to paleomagnetic information, some steps are necessary:

- averaging the direction of ChRM of the specimens from a same site;
- evaluate whether this direction does reasonably represent the direction the Earth's field had when the rocks were formed;
- averaging the paleomagnetic directions of different sites of the same geological unit;
- comparing the directions of different geological units.

4.4.1
Statistical Analysis

Having N ChRM directions (D_i, I_i) of specimens from the same site, the first step is to compute their average value, obtained through the vector sum of the N unit vectors. In the geographic reference system (x, y, z = north, east, down), the three direction cosines (l_i, m_i, n_i) of each vector are

$$l_i = \cos D_i \cos I_i; \quad m_i = \sin D_i \cos I_i; \quad n_i = \sin I_i$$

The magnitude of their resultant (Fig. 4.20) is $R^2 = (\Sigma_i l_i)^2 + (\Sigma_i m_i)^2 + (\Sigma_i n_i)^2$, and its direction cosines are

$$l = (\Sigma_i l_i) / R; \quad m = (\Sigma_i m_i) / R; \quad n = (\Sigma_i n_i) / R$$

Declination and inclination (D, I) of the site mean ChRM direction are given by

$$D = \arctan(m / l); \quad I = \arcsin(n)$$

In paleomagnetism, the significance of the mean direction, or – which is the same – the dispersion of the data from which it was calculated, is evaluated using the Fisher's statistics. Each direction corresponds to a point on a unitary sphere and assuming they are distributed with azimuthal symmetry about the "true" value, the probability density function $P(\theta)$ of Fisher's distribution is given by

$$P(\theta) = k \exp(k \cos\theta) / 4\pi \sinh(k)$$

where θ is the angle between an individual direction and the "true" one and k the precision parameter, which gives the concentration of the points and varies from $k = 0$ if distribution is uniform, i.e. the directions are randomly dispersed, to $k \to \infty$ when they are concentrated near the "true" direction. In reality, we do not know the "true" direction and the value of k; so we can only calculate their best estimates, which are the mean direction, corresponding to the resultant, and the precision $k = (N - 1) / (N - R)$. The "true" direction has a certain probability of falling within a cone centered on the mean and with apical semi-angle α, which in the case of Fisher's statistics is given by the relationship

$$\cos\alpha_{(1-p)} = 1 - (N - R) / R \cdot ((1 / p)^{1/(N-1)} - 1)$$

where $(1-p)$ is the level of confidence. Usually, one sets $p = 0.05$, so the angle α_{95} corresponds to the half-angle of the cone within which the true direction lies at the 95% probability level. Figure 4.21 shows some real cases. The angle α_{95} can be considered as the error that affects the mean direction; the corresponding errors for inclination and declination are given by

$$dI = \alpha_{95}; \quad dD = \alpha_{95} / \cos I$$

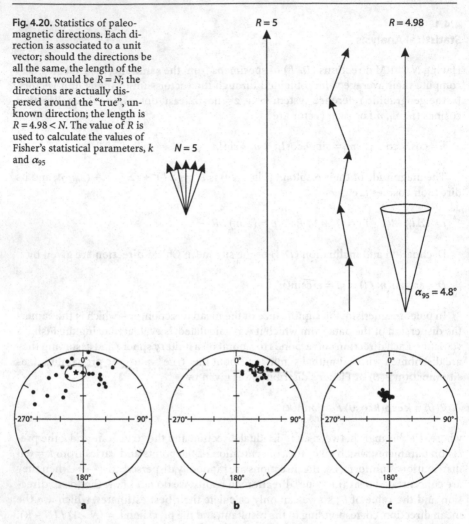

Fig. 4.20. Statistics of paleo-magnetic directions. Each direction is associated to a unit vector; should the directions be all the same, the length of the resultant would be $R = N$; the directions are actually dispersed around the "true", unknown direction; the length is $R = 4.98 < N$. The value of R is used to calculate the values of Fisher's statistical parameters, k and α_{95}

Fig. 4.21. Equal-area projection of paleomagnetic directions from single sites. Symbols: *dots* = ChRM direction; *star* = site mean direction with ellipse of confidence; **a** high dispersion ($\alpha_{95} = 11.5°$, $k = 7$ – Miocene marls, Ligurian-Piedmontese Basin, NW Italy); **b** low dispersion ($\alpha_{95} = 4.5°$, $k = 43$ – Jurassic limestones, *Rosso Ammonitico*, western Sicily); **c** very-low dispersion ($\alpha_{95} = 2.1°$, $k = 423$ – fine-grained pyroclastics, Vesuvius, Italy)

In the stereographic net, the cone of confidence is projected as a circle (equal-angle, or Wulff net) or an ellipse (equal-area, or Schmidt net).

The numerical values of the various statistical parameters are useful to give a quantitative evaluation of dispersion, but they must not be considered too rigidly. Each statistic presupposes a high number of data and their mutual independence, i.e. that they correspond to a random choice within the population to be analyzed. In paleomagnetic studies, oftentimes the data are not very numerous (there are rarely more than 10–15 specimens measured for a site) and randomness depends on the way sampling was done, which may not be random. Figure 4.22 shows the case of a lava flow from Mount Etna, sampled at 10 cm interval along its whole thickness of about 2 m. The mean direction is very close to that of the Earth's field (known, because the flow dates back to 1971) although declination and inclination have variations of nearly 20°. The within flow variations are likely to be random and the good sampling strategy minimizes their effects. A sampling concentrated in one part of the flow would have lost the randomness: data dispersion would probably have been minor, but the mean direction would have been systematically deviated relative to the field.

The values of the statistical parameters are essential to evaluate the quality of the paleomagnetic data, but the weight to be attributed to the paleomagnetic result depends first of all on the logic of the sampling. For example, in the case of an outcrop of sedimentary rocks sampling an individual layer gives different results than sampling various layers. In the first case, the site mean direction will correspond to a short time interval and have a low dispersion, in the second the dispersion will be higher, since the corresponding time interval is longer.

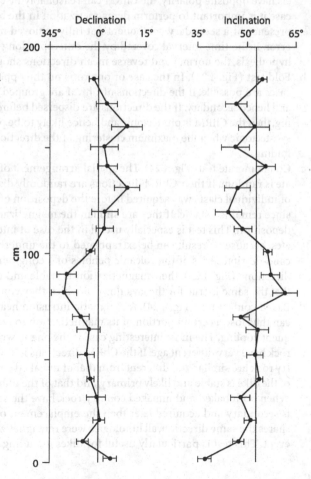

Fig. 4.22. Changes in declination and inclination of TRM throughout an individual lava flow (1971 eruption of Etna). *Vertical lines* show the *D, I* values of the Earth's magnetic field at Etna in year 1971 (modified after Rolph 1997)

Other statistical procedures apply to particular problems, such as determining whether two mean directions are different from each other or giving a quantitative evaluation to the tests described in the next section.

4.4.2
Field Tests

Once the site mean direction of ChRM is determined, the paleomagnetist has to give it a geological meaning. The process is typically deductive and has to considers both the magnetic and the geological data. The field tests are similar criteria to those typical of stratigraphy and can provide indications on the primary nature of the ChRM. Traditional tests are:

a Reversal test (Fig. 4.23). In the case of sedimentary rocks or superposed lava flows, the time elapsed during the deposition of the layers or the emplacement of the flows may cover one or more polarity reversals. If the specimens collected at various levels have opposite polarity, the ChRM can reasonably be considered primary. In this case it is important to perform demagnetization in the best possible way, since the presence of a secondary component not fully removed may introduce a systematic error. If the time interval covered by the outcrop is long enough to apply the GAD hypothesis, the normal and reverse mean directions should be antipodal.

b Fold test (Fig. 4.24). In the case of outcrops on the opposite limbs of a fold, three cases are possible. If the directions of ChRM are grouped, the ChRM is post-tectonic and hence secondary. If the directions are dispersed before and grouped after unfolding, then the ChRM is pre-tectonic and hence likely to be primary. Lastly, the ChRM is syn-tectonic when the maximum clustering of the directions occurs after partial unfolding.

c Conglomerate test (Fig. 4.24). The spatial arrangement of the clasts of a conglomerate is random. If their ChRM directions are randomly dispersed, the magnetization of individual clasts was acquired before the deposition of the conglomerate and has since remained stable. If they are similar, the magnetization was acquired after the deposition. This test is especially useful in the case of inter-formational conglomerates, because its result can be extrapolated to the underlying source rock. In a volcano-detrital series, if the volcanic pebbles of a conglomerate have random ChRM directions (Fig. 4.25), their magnetization is stable and it is reasonable to suppose that the same is true for the lava flows of which they come from.

d Baked contact test (Fig. 4.24). A magmatic intrusion heats the country rock, which can thus lose a certain portion of its magnetization to reacquire it during its subsequent cooling. The most interesting case is the one in which a dike and the country rock have very different age. If the ChRM directions in the dike and the heated country rock are similar and different from that of unbaked country rock, then the ChRM of the dike is stable and likely primary, and that of the unbaked country rock is older. When both baked and unbaked country rock have the same direction, their ChRM is secondary and acquired later than the emplacement of the dike. If even the dike shares the same direction, all lithologies were remagnetized during a major regional event. This test is particularly useful for dikes intruding a crystalline basement.

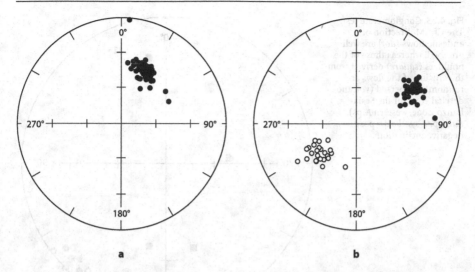

Fig. 4.23. Reversal test. Equal-area projection of NRM (**a**) and ChRM (**b**) directions (Bajocian limestones, Betic Cordillera, Spain). Symbols: *full/open dots* = positive/negative inclination. All NRM directions have normal polarity; ChRM directions after thermal demagnetization have both normal and reverse polarity. A Tertiary overprint hides the primary magnetization, which shows a succession of 5 normal and 6 reverse polarity intervals recorded in a 3-meter-thick section (courtesy C. De Giorgis)

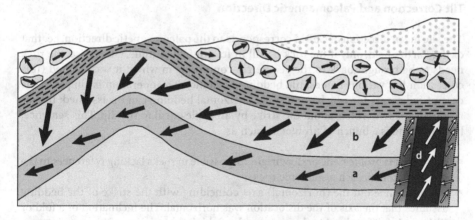

Fig. 4.24. Paleomagnetic field tests for stability. (1) Fold test. Remanence direction varies along layer *b*, and is uniform along layer *a*. Unfolding makes uniform the directions in *b*; remanence predates folding and may be primary in origin. Remanence in *a* is secondary, acquired after folding. (2) Conglomerate test. The dispersion of the remanence direction of pebbles in layer *c*, which originated from rocks of layer *b*, strenghtens the hypothesis of primary remanence in layer *b*. (3) Baked contact test. The dike *d* intruded during a reverse polarity interval and remagnetized the country rock close to its walls. Rocks of layer *b* still have the same direction of the dike, whose remanence can be inferred to be primary. Rocks of layer *a* have the same direction they have far from the dike; the remagnetization of layer *a* occurred during a normal polarity interval later than the dike intrusion

Fig. 4.25. Conglomerate test.
The ChRM direction of
andesite flows (*dot*) are well
grouped, whereas those of the
boulders (*square*) derived from
the erosion of the flows are
randomly dispersed (volcano-
detrital cover of the Sesia-
Lanzo Zone, western Alps).
Full/open symbols = positive/
negative inclination

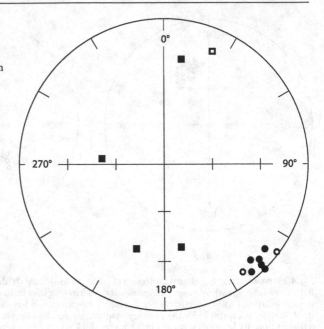

4.4.3
Tilt Correction and Paleomagnetic Direction

The site mean direction of ChRM corresponds to the paleomagnetic direction, i.e. that of the Earth's field, only if the rock is still in its original attitude. Otherwise, the rock needs to be brought back to the geometric conditions in which it was formed (this correction has various names: tilt, bedding, tectonic). The operation is simple in the case of sediments for which an original horizontal bedding can be assumed: the direction is made to rotate about the strike by an angle equal to the dip. This sentence, however, entails a bunch of problems, such as

- correction is problematic or downright impossible in rocks lacking references to the paleohorizontal, such as plutonic rocks;
- the rotation about the (horizontal) axis coinciding with the strike of the bedding assumes that the axis of the dislocation was horizontal. The inclination of a fold or tilting axis is often hard to derive from the field data;
- if the dislocation is multi-phase, the chronological sequence must be known since the total correction depends on the order of the individual corrections.

Once the paleomagnetic directions of the individual sites of a geological unit are obtained, the paleomagnetic direction of the unit (*D, I*) is their mean direction, calculated using Fisher's statistic as before. If attitude changes from one outcrop to the other, the quality of the tilt correction can be evaluated by comparing the dispersion of the mean direction before and after the correction, as in the fold test.

4.4.4
Virtual Geomagnetic Pole (VGP) and Paleopole

Since the direction of the magnetic field along the Earth's surface depends on geographic coordinates, the paleomagnetic directions of distinct geological units can be compared directly only on a regional scale. On a broader or planetary scale, geographic dependence must be eliminated and a common reference must be used. As a first approximation, the Earth's field is modeled as a geocentric axial dipole (GAD). In this hypothesis, the direction of the field will always concur to the pole, regardless of the position of the point taken in consideration. To compare the data of different geological units, therefore, it is necessary to go from the paleomagnetic direction to the virtual geomagnetic pole (VGP). Given a direction (D, I) obtained in the site S (of geographic coordinates λ_S, φ_S), the corresponding VGP is along the great circle that passes through S (Fig. 4.26) and forms, with the geographic meridian an angle equal to the declination D. The angular distance between S and the VGP is the colatitude p, which is calculated with the dipole formula $\tan I = 2 \tan \lambda = 2 \cot p$ (i.e. Eq. 1.42 written using for latitude the symbol λ instead of Θ)

$$p = \mathrm{arcot}(1/2 \tan I) \qquad (4.11)$$

Latitude and longitude (λ_P, φ_P) of the VGP are given by the formulas

$$\lambda_P = \arcsin(\sin \lambda_S \cos p + \cos \lambda_S \sin p \cos D)$$

Fig. 4.26. Calculation of the VGP position: D = site mean declination; p = site colatitude (see Eq. 4.11); dp, dm = radii of the VGP ellipse of confidence

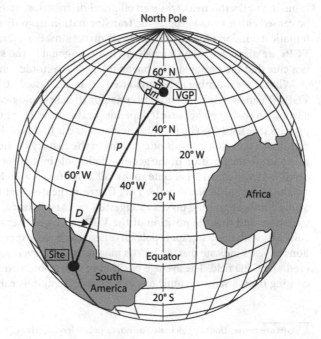

$$\varphi_P = \varphi_S + \beta \qquad \text{if } \cos p \geq \sin \lambda_S \sin \lambda_P$$

$$\varphi_P = \varphi_S + 180° - \beta \qquad \text{if } \cos p < \sin \lambda_S \sin \lambda_P$$

where $\beta = \arcsin(\sin p \sin D / \cos \lambda_P)$. The circle of confidence α_{95} associated to a mean direction is transformed for the VGP into an ellipse, because the relationship between latitude and inclination of the Earth's magnetic field is not linear. The error dI is translated into a colatitude error, which corresponds to the radius dp of the confidence ellipse along the great circle from S to the VGP; the error dD corresponds to the radius dm in the perpendicular direction (Fig. 4.26).

The non-linearity of the latitude-inclination relationship passing from direction to pole causes a methodological problem when analyzing sites of a same geological unit (or of geological units of the same age): should the mean direction be calculated first, and then the corresponding VGP, or should the VGPs of the individual sites be first calculated, and then their mean value? Remember first of all that the mean position of VGPs is calculated using Fisher's statistics again: longitude and latitude respectively take the place of declination and inclination, and the statistical parameters are conventionally designated with capital letters, K and A_{95}. The two possible methods give the same position as the mean VGP, to which is associated a confidence ellipse (dp, dm), if the mean of the directions is computed and then the VGP is calculated, or a circle (A_{95}), if the individual VGPs are calculated and then their mean is computed. The difference is in the different distribution: a circular distribution of directions produces an elliptical distribution of VGPs, and vice versa. This difference causes both statistical problems, since Fisher's statistics presupposes circular symmetry, and data interpretation problems. A circular distribution is considered an effect of random variations in the Earth's field, while an elliptical distribution, elongated in one direction, can be caused either by the geographic transformation from direction to pole or by a systematic deviation of the paleomagnetic directions. For example, the case in which the VGPs are dispersed along a small circle, orthogonal to the site to pole paleomeridian, is a clue that the sampled sites have undergone tectonic rotations about vertical axes.

What is the meaning to be attributed to an individual VGP and to a mean of a set of VGPs? To answer, it is essential to consider the chronological aspect, which differs according to rock type because it depends on the duration of the remanence acquisition process (Sect. 4.1.7). The most immediate case is that of volcanic rocks. The magnetization of a lava flow records one instant in the history of the Earth's field: its VGP corresponds to the position of the geomagnetic pole in that same instant. Nearly all VGPs obtained from Plio-Pleistocene rocks fall at latitudes $>70°$ N[2] within the cap centered on the geographic North Pole. This figure indicates that the magnetic pole moves around the geographic pole, as suggested by the historical measurements of secular variation, and that the position of the VGPs averaged over times in the order of 1 Myr coincides with the geographic pole. In other words, the experimental data of the last millions of years indicate that the Earth's magnetic field, averaged over time, can be considered as a GAD field. The average position of VGPs obtained from rocks of similar age coming from a same tectonic or geodynamic domain is called a *paleomagnetic pole*.

---

[2] VGPs are conventionally calculated as north poles, irrespective of the ChRM polarity.

4.5
Paleomagnetic Information

The remanent magnetization of rocks contains a good deal of information on physical processess, geometry and time. This section provides examples relating to geodynamics, structural geology and geophysics, while Chap. 7 is dedicated to the use of paleomagnetism as a relative chronology tool.

4.5.1
Geodynamics and Paleogeography

In the previous section we have seen that the GAD hypothesis is substantially valid for the last 5 Myr. Paleomagnetic analysis of older rocks shows on the one hand that VGP dispersion over times of some Myr is similar to the one observed for Plio-Pleistocene, on the other hand that their average position, i.e. the paleomagnetic pole, is distinct from the present-day geographic pole. In the case of Europe (Fig. 4.27), for example, Permian VGPs are in the easternmost part of the Pacific Ocean and are distributed astride the 45° N parallel. Europe's Permian paleomagnetic pole is located 4 000–5 000 km away from the geographic pole. More in general, the position of the paleomagnetic poles from the various continents varies in time and they move farther away from the present-day geographic pole the greater the age of the rocks. The curve drawn joining the paleomagnetic poles is called apparent polar wander (APW) path. The size of the movements (Fig. 4.28) indicates that their cause is a geophysical phenomenon of first magnitude, and two different hypotheses can be put forward:

Fig. 4.27. European paleomagnetic poles from Pliocene-Pleistocene (*square*) and Permian (*dot*) rocks (from Lowrie 1997)

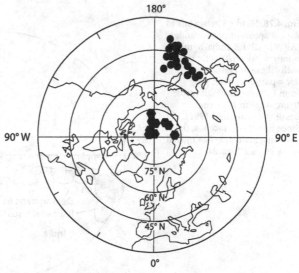

European paleomagnetic poles:
● Pliocene and Pleistocene
● Permian

- the magnetic pole has moved along the surface of the Earth; in other words, the axis of the dipole is not bound to the axis of rotation and the angle between the two axes can take on any value;
- the magnetic pole has remained fixed and the continents have moved: the two axes have remained substantially co-linear and the continental masses have been the ones that moved.

The observation that different continents have different APW paths excludes the first hypothesis, since an absolute movement of the pole would be recorded the same way in any point of the Earth's surface, and lends plausibility to the second hypothesis. This hypothesis found the first confirmation in the 1950–1960 decade as a result of the systematic study of Palaeozoic and Mesozoic rocks of Europe and North America. The APW curves of the two continents are clearly different (Fig. 4.29), however, their shape is similar and a rotation of 38° brings the two curves to overlap. This rotation corresponds to the one necessary to close the Atlantic Ocean up according to the fit of the coastlines proposed by Bullard, Everett and Smith in 1965. Paleomagnetism thus provides a proof of the movements of the continents (or rather of the plates) that is independent of any other geodynamic assumption and based only on the hypothesis that the Earth's magnetic field has remained essentially dipolar over geological times. The whole of paleomagnetic data accumulated over fifty years supports the validity of the hypothesis and is fully compatible with geological and geophysical observations that suggest, but do not prove, relative movements between the lithospheric plates. The paleogeographic reconstructions of the Pangea supercontinent and of its northern and southern parts, Laurasia and Gondwana, hypothesized during the first half of the 20th century, mainly by Wegener and DuToit, have found substantial confirmation, although some aspects are still controversial to this day.

Fig. 4.28. Middle Cretaceous to Recent apparent polar wander (APW) path for India. Symbols: *square* = paleomagnetic pole with ellipse of confidence; *dot* = paleomagnetic pole from Cretaceous basalt traps; *figure* = age in Ma (from Besse and Courtillot 1991). Since middle Cretaceous, the paleomagnetic pole has moved some 90° northwards

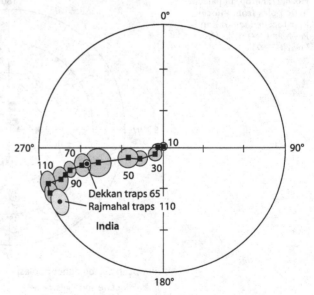

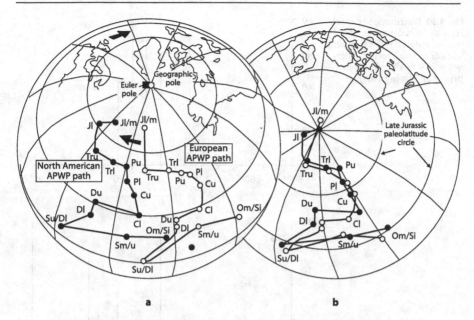

Fig. 4.29. North American and European APW paths from middle Ordovician/lower Silurian (Om/Sl) to late-middle Jurassic (Jl/m); **a** APW paths in present-day geographic coordinates; **b** APW paths (late Jurassic geographic coordinates) after rotating Europe about the Euler pole in *a* (from Lowrie 1997; data after Van der Voo 1990)

The paleogeographic evolution of the Earth's surface from Mesozoic to Recent is quite well defined, thanks to the integration between the continents' APW paths, the sequence of the ocean-floor magnetic anomalies (Chap. 7) and the constraint of coastal contour fitting. Delving farther back in time, things become more complicated (e.g. there are three models for the reconstruction of Pangea: A, A2, B) and ultimately speculative, like the connections between the various Archaeozoic cratons.

Paleomagnetic data allow to reconstruct the relative motions of two continents, but they provide no information on their absolute position on the Earth's surface. Paleolatitude can be derived, because it is a function of the inclination of the paleomagnetic direction, but paleolongitude remains indeterminate, since the GAD model has the Earth's axis of rotation as its axis of symmetry. Latitude is the main variable that affects climate and therefore it is possible to correlate paleomagnetic data with paleoclimatic data, which can be deduced from the fact that certain types of rocks and many fossils have a geographic distribution limited to certain climatic regions. The comparison cannot be very quantitative, but qualitative concordance between the results of completely different methodologies is always a convincing argument. A great variety of rocks are formed under particular climatic conditions: evaporites, red beds, coal, coral reefs. The latter are a classic example: for the development of a reef to take place, sea water temperature must not drop below 22 °C, a condition that is met only in the 30° S–30° N latitude band. The distribution of present-day reefs versus latitude is symmetrical with respect to the equator (Fig. 4.30), whereas that of fossil reefs is

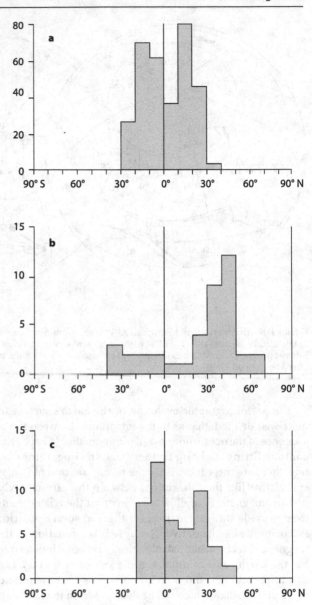

Fig. 4.30. Distribution of coral reefs vs. latitude; **a** present latitude of modern reefs; **b** present latitude of fossil reefs; **c** paleo-latitude of fossil reefs (from Briden and Irving 1964)

strongly asymmetrical and many are located at higher latitudes than 30°, in particular in the northern hemisphere, where a frequency peak is recorded between 30° N and 50° N. Bringing the reefs back to their original latitude deduced from paleomagnetism, distribution becomes more symmetrical and 95% of the reefs fall within the 30° S–30° N latitude band.

The paleomagnetic correlations that confirm an equatorial symmetry of paleoclimatic indicators imply an axial dipole and are therefore a very strong argument in favor of the hypothesis that the GAD model has remained substantially valid over geological epochs. Other connections between magnetic field, rock magnetism and climate will be discussed in Chap. 8.

The analysis of paleomagnetic directions, or VGPs, is a well consolidated application of paleomagnetism to geodynamics, but other information can be derived from analyzing the ways in which the remanent magnetization was acquired. A recent example is thermopaleomagnetism, which aims to evaluate the cooling of rocks connected to vertical tectonic movements. Its first systematic application was performed on Liassic marly limestones and black shales of the Dauphinoise Zone (western Alps), metamorphosed during the upper Oligocene. In the prograde phase of metamorphism, whose thermal peak was about 350 °C, pyrrhotite grains were formed, which did not acquire a remanent magnetization because the Curie point of pyrrhotite is $T_c \approx 320$ °C (Sect. 2.3.3). The subsequent cooling was caused by the exhumation of the chain, and hence it was very slow, in the order of a few °C/Myr. In such a case, the acquisition of TRM extends over a period of time that is comparable with the succession of the polarity reversals and the rock records two antipodal directions. If the spectrum of the blocking temperatures T_b is sufficiently broad (≈ 100 °C), the rock records many polarity reversals and its TRM is made up by many components (PTRM) with opposite directions. The curve of the TRM intensity versus temperature, therefore, does not have a monotonous growing profile as in Fig. 4.2, but it is a broken line that alternates segments with opposite slope (Fig. 4.31). If the PTRM acquired immediately below the Curie point has normal direction, the one acquired when polarity has changed has reverse direction and hence is subtracted from the first one and the cumulative curve of the TRM decreases. At the subsequent polarity change, the new PTRM is parallel to the first one and the cumulative curve starts increasing again and so on, with a slope change in correspondence with every reversal. If the remanence of the rock is exclu-

Fig. 4.31. Thermal demagnetization of laboratory-controlled TRM for SD pyrrhotite grains. Symbols: *tTRM* = "total" TRM acquired under constant-polarity field; *cTRM* = "composite" TRM including five polarity reversals (see text for further explanation) (from Crouzet et al. 2001)

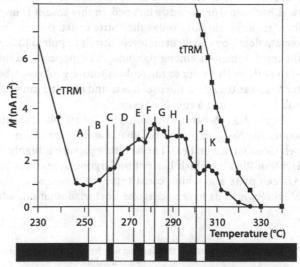

Polarity of the field during the creation of the composite TRM

sively of thermal origin and the pyrrhotite grains are in the SD state, the process is reversible (Thellier law) and a highly detailed thermal demagnetization allows to reconstruct the curve of the PTRM, from which the sequence of polarity reversals can be deduced.

This phenomenon is wholly similar to the magnetization of the oceanic crust (Sect. 6.2); only the time factor is recorded differently: in the oceanic crust, each polarity period corresponds to a certain width of the anomaly band; in the case of thermopaleomagnetism it corresponds to an interval in the T_b spectrum of the grains. Once the sequence of polarity reversals is obtained, the comparison with the GPTS reference scale (Sect. 7.1) allows to evaluate its duration, and hence the cooling rate, from which the chain's exhumation rate can be estimated.

4.5.2
Regional Tectonics

Use of paleomagnetism in regional tectonics depends on the scale of the problem. In the case of a crustal block or a terrane, we have a tectonic domain which separated from a major plate and for a certain period of time moved independently on the Earth's surface. This movement has to be fitted in the global frame of the plates' motion and the paleomagnetic directions or the VGPs are compared with *expected* directions or VGPs, where expected means that they are calculated starting from an APW curve used as a reference on the grounds of a certain geodynamic hypothesis. Limiting ourselves to the directions, we see that an inclination I different from the expected one indicates a latitudinal movement (Fig. 4.32), while a discordance of the declination D indicates a rotation about a vertical axis. If both D and I are different from the expected ones, the movement of the block can be interpreted as a rotation around a point of the Earth's surface external to the block, called Euler pole[3].

At the local scale, i.e. within a same tectonic domain, the differences between the paleomagnetic directions are interpreted in terms of rotations, whose axes can be vertical, horizontal or variously inclined. In this case, it is important to keep in mind that local rotations do not involve the entire crust: the moving tectonic unit is freed at a certain depth by decollement levels. Studies applied to regional tectonics are perhaps the most numerous among paleomagnetic papers and the reader can seek in the literature those that refer to the regions whose geology (s)he knows personally, the better to assess the sense, the usefulness and also the limits of paleomagnetic results. We will just comment a couple of examples.

Figure 4.33 shows the paleomagnetic data obtained from Mesozoic pelagic limestones and Neogene limestones, chalks and clayey sediments of the Sicilian Maghrebide belt (Mediterranean Sea). The resulting picture is highly complex. With respect to the Hyblean Plateau, which has not undergone significant rotations with respect to the African plate and which constitutes the foreland of the Sicilian belt, the various Maghrebide units have undergone clockwise rotations which decrease from the most

---

[3] Euler's theorem states that the simplest way to connect two points on the surface of a sphere (in our case the initial and final location of a crustal block) is a rotation about an opportune vertical axis, whose pole is given by its intersection with the sphere.

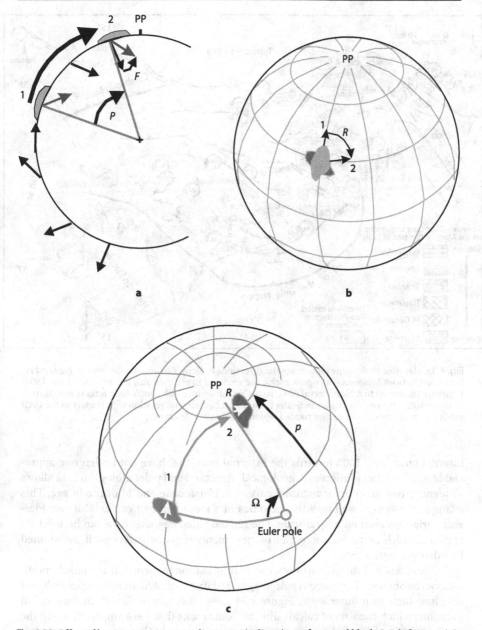

Fig. 4.32. Effect of large movements on paleomagnetic directions of a crustal block. Symbols: *1* = original position; *2* = present-day position; *PP* = paleomagnetic pole. **a** Latitudinal movement. Symbols: *black arrow* = direction of a dipolar field with magnetic pole at PP; *gray arrow* = paleomagnetic direction. The block was magnetized along the direction of the field (the black and gray arrows do coincide). Poleward translation of the block by an angle *P* toward PP results in a paleomagnetic direction shallower than expected by an angle *F*. **b** Rotation around a vertical axis. The block was magnetized along the paleomeridian (*Arrow 1*, pointing toward the pole); rotation has caused the paleomagnetic declination to rotate clockwise by the angle *R* (*Arrow 2*). **c** Rotation around an Euler pole. Rotation by an angle *Ω* results in a poleward translation *p* and a rotation *R* (from Butler 1992)

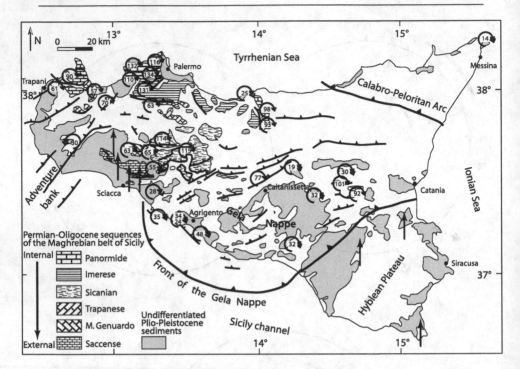

Fig. 4.33. Mesozoic to Neogene paleomagnetic directions in Sicily. *Circular arrows with single head* refer to results from Mesozoic-Paleogene rocks; the *enclosed figure* indicates the amount of clockwise rotation with respect to African coeval directions. *Circular arrows with double head* refer to results from Neogene rocks; the *enclosed figure* indicates the amount of clockwise rotation with respect to the GAD model. *Vertical arrows* refer to non rotated regions (courtesy F. Speranza)

internal units, over 100°, towards the external ones, that have not undergone appreciable rotations. Integration with geological-structural and paleontological data allows to identify two main deformational episodes, Pleistocene and Miocene in age. This example shows very well how the tectonics of a mountain belt can exhibit considerable variations even on a local scale. The related paleomagnetic data can be used for regional models or the reconstruction of movements of major plates only if constrained by other geological data.

In the absence of such constraints, useful information to formulate tectonic hypotheses can be obtained if numerous paleomagnetic data distributed over a sufficiently broad area have their own inner logic. Figure 4.34 shows the case of the North Patagonian batholith, which consists of calcalkaline plutonic rocks that were emplaced along the Pacific margin of South America in a succession of mainly Cretaceous intrusive episodes. Interpreting data from batholithic rocks is made difficult by the lack of a reference to paleohorizontal, and hence of evidence for or against possible tilt subsequent to magnetization. Distribution of the VGPs of the North Patagonian batholith exhibits considerable dispersion, but a careful inspection suggests that its shape is not random, but rather that the VGPs fall along a small circle, whose center is not far from the sampling region. The shape of the distribution can be explained as the result of rota-

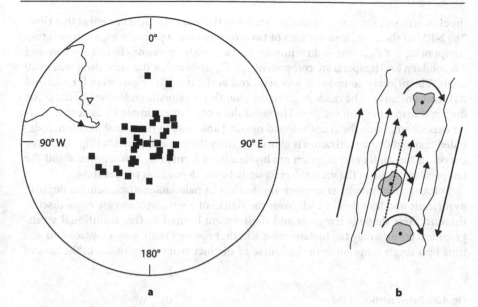

Fig. 4.34. Paleomagnetic results from the North Patagonian batholith (Chile); **a** equal-area projection of VGPs. Symbols: *square* = VGP; *full triangle* = sampling site; *open triangle* = center of the small circle interpolating the VGPs; **b** shear-zone model to explain the elongated distribution of VGPs. Rotation of crustal blocks around vertical axes moves their VGPs along a small circle centered in the sampling region (from Beck et al. 2000)

tions about vertical axes. These rotations change the value of declination, but not that of inclination: the angular distance of the VGPs from the sampling site therefore does not change and defines the small circle along which the VGPs fall. Distribution of VGPs thus suggests the presence of an extensive shear-zone parallel to the continental margin.

4.5.3
Volcanism

In the case of volcanic rocks, paleomagnetic study provides information both about the magnetic field of the past and on the remanence acquisition processes. Magnetization is a TRM and thus contains information about thermal conditions during emplacement. In the case of active volcanoes, the capability of estimating these temperatures is highly important because it provides a quantitative figure, useful in modeling expected eruptions and in assessing volcanic hazard. The most typical case is that of pyroclastic flows, for which the final deposition T can be estimated analyzing the NRM in the lithic clasts, that is fragments of a more ancient rock embedded in the deposit. They acquired their own primary remanence when the parent rocks formed, well before the emplacement of the pyroclastic deposit. If a flow picks up a clast at ambient temperature, the clast is heated to a temperature approaching the deposition temperature of the pyroclastic material, T_{dep}, and hence its ferromagnetic grains with $T_{\text{b}} \leq T_{\text{dep}}$ are demagnetized. While cooling after deposition, the grains are remagnetized and the clast acquires a remanence that is parallel to the magnetic field present at that time. The NRM of the clasts thus consists of two components: a primary, high-temperature component, $T > T_{dep}$, whose direction varies randomly from one clast to another and a secondary, low-temperature component $T \leq T_{dep}$, which has the same direction in all clasts. The primary component was acquired at the time the clasts were formed and its direction can only be random, given the chaotic movements undergone by the clasts during transport within the flow. The secondary component instead was acquired after deposition, when the clasts stopped moving and were in their final position. A detailed thermal demagnetization is able to identify the two components (Fig. 4.35): the curves in the Zijderveld diagram are broken lines formed by two segments and the temperature value at the vertex corresponds to the deposition temperature.

Volcanic stratigraphy is another application of paleomagnetism. Scoriae deposit, pyroclastic and lava flows which cover the flanks of a stratovolcano are often discontinuous, their shape is irregular and their extent limited so that traditional stratigraphical criteria may fail to determine whether distinct units were emplaced all at a time by a single eruption or in the course of distinct eruptive episodes. In the case of

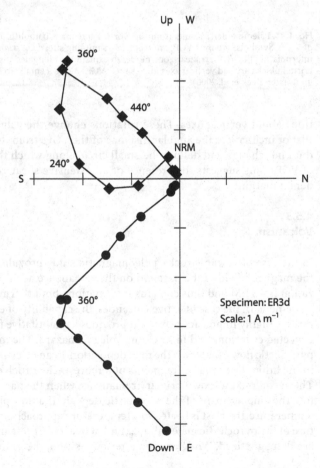

Fig. 4.35. Determination of the re-heating temperature by means of thermal stepwise demagnetization (lithic clast from pyroclastic deposits, Vesuvius, Italy). The low T_b components are completely erased at 360 °C

active volcanoes, the ability to ascribe various deposits to a single eruption helps in assessing the total volumes of the products and hence estimating the dimensions of the expected, future eruptions. If distinct volcanic units have the same paleomagnetic direction, it is likely that all of them share the same age; if they form two or more groups with different directions, then they were emplaced by eruptions occurred at different times.

4.5.4
Prospection

Magnetic anomalies are caused by the total magnetization of the rocks $J_{tot} = J_i + J_r$ and the role of the two types of magnetization depends on the Königsberger ratio $Q = J_r / J_i$ (Sect. 3.3). When $Q < 1$, induced magnetization J_i prevails and it is sufficient to know susceptibility κ to calculate the model of the source bodies. In the case of volcanic rocks, $Q > 1$ and the contribution of remanent magnetization J_r is no longer negligible; in fact, in many cases it is preponderant. The most typical case is that of sea-floor magnetic anomalies, which we will examine in detail in Chap. 6. In this section we discuss two case histories from Italy's aeromagnetic map.

An aeromagnetic survey was performed during geothermal investigations in the Vico-Cimini volcanic district in central Italy. Figure 4.36 shows the intensity of magnetization of various lithological types of the two volcanoes: it is evident that $J_r \gg J_i$. The most interesting thing, however, is observed along the profile of magnetic anomaly reduced to the pole (Sect. 3.3.4): in correspondence with the volcanic structures surrounding the Vico lake, there are two positive anomalies, while in the Cimini region the anomaly is negative. The reason for this difference is the different age of the two volcanic complexes: 0.4–0.14 Ma for Vico, 1.35–0.95 Ma for the Cimini. The rocks of the former acquired their remanence during the normal polarity Brunhes epoch (Chap. 7): J_i and J_r have the same direction and sum up. The rocks of the latter acquired their remanence during the reverse polarity Matuyama epoch. Their strong J_r has reverse direction and completely masks J_i. The total magnetization of the Cimini rocks is directed upwards and, being in the northern hemisphere, the anomaly is negative. For a correct interpretation of the anomalies, not only polarity is important, but also the J_r intensity. The anomaly depends on the magnetic moment of the source body, $M = VJ_{tot}$, and neglecting the contribution of remanent magnetization inevitably entails an overestimation of the volume.

In the region around Macomer (northwestern Sardinia, Mediterranean Sea), extensive tabular effusions of Plio-Pleistocene basalts overlie Oligo-Miocene calcalkaline volcanites, which in turn overlie a Palaeozoic crystalline basement. This region is characterized by a large magnetic anomaly whose maximum-minimum axis is oriented NW-SE (Fig. 4.37), unlike the typical N-S direction of the Italian region, where the declination of the Earth's magnetic field is close to 0°. Interpretation models show that a good fit between the measured and calculated anomaly requires the source body be formed by two structures with different magnetization directions: $D = 0°$, $I = 50°$ for the more superficial structure, $D = 330°$, $I = 50°–55°$ for the deeper one. These figures are consistent with the ages of the rocks and their paleomagnetic directions. The Oligo-Miocene volcanites were emplaced before the final phase of counterclockwise rotation of Sardinia and have TRM directions directed, on average, 30° to the northwest with respect to the present field. They are the main source of the anomaly and their

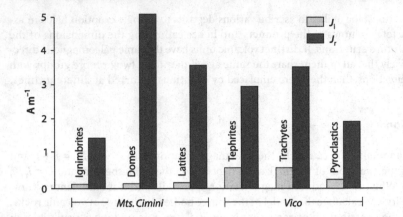

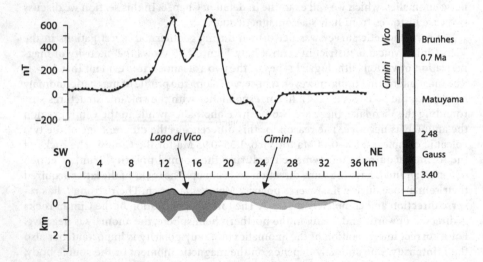

Fig. 4.36. Vico-Cimini volcanic district (Latium, central Italy); **a** induced (*gray*) and remanent (*black*) magnetization of the main lithologies; **b** magnetic anomaly reduced to the pole and interpretation model (see text for further explanation) (modified after Gandino et al. 1989)

remanence produces the deviation of its axis. On the other hand, the Plio-Pleistocene basalts have TRM directions close to the north: their remanence enhances the anomaly with limited effect on its axis.

4.5.5
Paleofield

A different approach to the paleomagnetic data consists in retrieving information about the physical properties of the Earth's field in the past, and hence to contribute to the understanding of its complex phenomenology and the improving of the models.

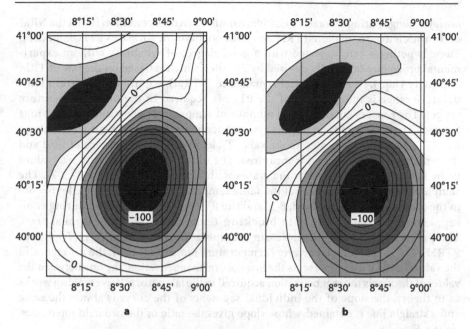

Fig. 4.37. Aeromagnetic anomaly map of NW Sardinia (contour interval = 10 nT); **a** measured anomaly; **b** computed anomaly (see text for further explanation). The maps are upward continued to 20 km, to reduce the effect of surface rocks (from Galdéano and Ciminale 1987)

A first point is the paleointensity of the field, which allows (Eq. 1.7) to derive the magnetic moment of the GAD, called virtual axial dipole moment (VADM). The determination of paleointensity presupposes the knowledge of the relationship between the intensity of the remanence acquired by a rock, J_r, and the intensity (actually the magnitude) of the magnetic field F that caused it. In the case of the TRM, the relationship is based on Néel's theory of single domain grains and can be expressed in a simple form. The argument of the tanh function in Eq. 4.3, $a = \mu_0 V J_s(T_b) H_E / K T_b$, is $a \ll 1$ for typical values of its parameters; its expansion in series can thus be limited to the first term, then $\tanh(a) \approx a$ and one obtains the relationship

$$J_{TRM}(20\,°C) \approx N J_s(20\,°C)\, a = N J_s(20\,°C) \mu_0 V J_s(T_b) H_E / K T_b$$

which is linear in $\mu_0 H_E = F$, and can be rewritten as $J_{paleo} = C F_{paleo}$ where C is a constant that combines all the terms that depend on the characteristics of the ferromagnetic grains. It is therefore clear that if the rock is demagnetized thermally and then cooled in the presence of a known field B_{lab}, the similar relationship $J_{lab} = C B_{lab}$ applies. Combining the two relationships, C is eliminated and the paleointensity is given by

$$F_{paleo} = (J_{paleo} / J_{lab}) B_{lab}$$

For the method to be valid, C must actually be constant: the properties of the grains must not have changed since the rock was formed and acquired its TRM, and must noways change during laboratory heating. Another essential condition is that the NRM of the rock is a pure, unaltered TRM and does not have any CRM or VRM component. These hypotheses can be tested with a good degree of reliability with an experimental procedure originally devised by Thellier, which is based on the law of TRM additivity (Eq. 4.4): every partial remanence (PTRM) acquired in a certain temperature range is independent of the PTRMs acquired in the other temperature ranges. The process entails a good number of temperature steps and two heatings for each step. After measuring the remanence intensity at ambient temperature, $J_{paleo}(T_0)$, the sample is heated to the value T_1, let to cool in the absence of field and the remaining TRM, $J_{paleo}(T_1)$ is measured. The difference between these two values is the natural PTRM carried by the grains with blocking temperature $T_b \leq T_1$. The sample is heated a second time at the same temperature T_1, but this time let to cool in the presence of a known field, B_{lab}, and the PTRM ($T_1 - T_0$) acquired during cooling, clearly by the grains with blocking temperature $T_b \leq T_1$, is measured. The process is repeated for increasing temperature values and $J_{paleo}(T_i)$ versus PTRM ($T_i - T_0$) is plotted. For every temperature range (Fig. 4.38), the difference in the values of the y-axis represents the lost natural magnetization, the one between the values of the x-axis the magnetization acquired in the laboratory. If everything works as in theory, the slope of the individual segments of the curve is always the same and a straight line is obtained, whose slope gives the ratio of the two field intensities, $slope = F_{paleo} / B_{lab}$.

The intensity F of the Earth's field depends (Eq. 1.7) on the moment M of the dipole, the Earth's radius R and the latitude λ of the point where F is measured

Fig. 4.38. Determination of the paleointensity. For each temperature step, the ratio of the lost natural remanence (ΔNRM) to the acquired laboratory remanence (ΔPTRM) is constant

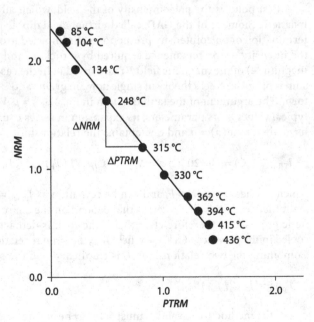

$$F = \frac{\mu_0}{4\pi} \frac{M}{R^3} \sqrt{3\sin^2 \lambda + 1}$$

Knowing F_{paleo}, therefore, the value of the moment (VADM) can be calculated. As usual, the greatest limitation of volcanic rocks is that of providing discontinuous information; to have continuity, it is necessary to use sediments. For these, there is no simple formula linking F and J_r, neither the process of acquiring the NRM can be repeated in the laboratory. However, a relative value of paleointensity can be obtained, giving an artificial magnetization to the rock, e.g. an ARM (Sect. 8.2). The intensity of the ARM depends on the content of ferromagnetic minerals in the rock, as does the intensity of the NRM. Normalizing the value of J_{NRM} of every sample with the value of the ratio of the intensities NRM/ARM, the dependence on the ferromagnetic content is eliminated, or at least strongly reduced, and only the dependence on F_{paleo} remains. The variations in J_{NRM}, for example along a core, yield therefore the variations in paleointensity, provided the other magnetic properties do not change, either in terms of minerals or of domains. Clearly, the sediments in which the sole or prevalent ferromagnetic mineral is SD magnetite are the most suitable.

Sediment analysis is very useful to have information both on the position of the VGP and on the intensity of the field during the transition from one polarity to the other. We know that reversals have occurred in the Earth's history, yet we do not know the way they occur and can only roughly estimate the time needed to complete the transition as a few kyr. In theory, many different models can account for a polarity reversal. Some models are schematically shown in Fig. 4.39:

a The moment of the dipole can first decrease and then grow in the opposite direction. The VGP would remain in its initial position as long as the intensity of the main dipole field is greater than that of the non-dipolar terms, then the field would be very weak and even disappear and finally the new VGP would go to the reverse position. The most dramatic consequence of this mechanism would be the vanishing of the field and thus of the shielding provided by the magnetosphere.
b The axis of the dipole can rotate without the moment necessarily changing. In this case the geographic latitude of the VGP would progressively decrease. The VGP would follow a path along a meridian, cross the equator and finally approach the opposite geographic pole.
c A quadru- or octupolar term can develop with opposite polarity to that of the main dipole and gradually "eat it".

By now it is generally accepted that the intensity of the Earth's field decreases by 50% and even more (Fig. 4.40) in correspondence with a polarity reversal or even with just an excursion, while opinions differ widely with regard to the directions. For now, paleomagnetic data do not yield clear indications. The hypothesis that transitional VGPs during the last millions of years have followed two antipodal preferential paths, one crossing the American continent, the other one Asia and Australia, has been at the center of a lively debate, but the question of a systematic behavior during the transition is still unresolved.

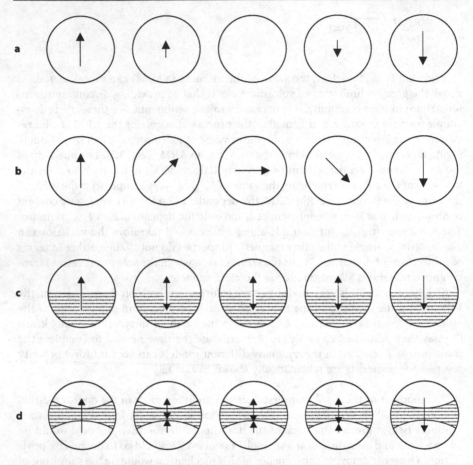

Fig. 4.39. Possible models for the Earth's field during a polarity transition (reverse to normal); **a** decrease in the axial dipole moment; **b** rotation of the main dipole without changes in moment; **c** quadrupole transition initiated in the southern hemisphere of the core; **d** octupole transition initiated in the low-latitude region of the core (from Merrill et al. 1996)

In this, as in other cases, use of paleomagnetic data to obtain physical characteristics of the field presents many challenges: irregular geographic distribution of the data, smoothing of the signal in sedimentary rocks, incomplete cleaning of magnetic overprints. For example, systematic deviations of the VGP, which are calculated with the GAD model, can be interpreted as the effect of second order contributions (quadru- and octupolar) that survived for long periods of time. There is good evidence, at least for the last 5 Myr, that these effects are not negligible, but statistical analysis is not yet able to provide sufficiently precise indications to constrain geodynamo models.

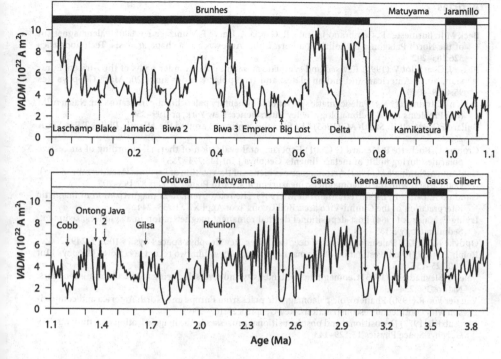

Fig. 4.40. Relative variations of the Earth's virtual axial dipole moment (VADM) during the last 4 Myr. Polarity reversals (*solid arrows*) and excursions (*open arrows*) correlate with VADM minima (from Valet and Meynadier 1993)

Suggested Readings and Sources of Figures

Books

Butler RF (1992) Paleomagnetism: Magnetic domains to geological terranes. Blackwell Scientific Publications, Oxford, UK, 319 pp

Collinson DW (1983) Methods in rock magnetism and palaeomagnetism. Chapman and Hall, London, 503 pp

Dunlop DJ, Özdemir Ö (1997) Rock magnetism. Fundamentals and frontiers. Cambridge University Press, Cambridge, UK, 573 pp

Lowrie W (1997) Fundamentals of Geophysics. Cambridge University Press, Cambridge, UK, 354 pp

Merrill RT, McElhinny MW (1983) The Earth's magnetic field. Academic Press, San Diego, California, USA, 401 pp

Merrill RT, McElhinny MW, McFadden PL (1996) The magnetic field of the Earth: Paleomagnetism, the core and the deep mantle, Academic Press, San Diego, California, USA, 531 pp

O'Reilly W (1984) Rock and mineral magnetism. Blackie, Glasgow, UK, 220 pp

Tarling DH (1983) Palaeomagnetism. Chapman and Hall, London, 379 pp

Tauxe L (1998) Paleomagnetic principles and practice. Kluwer Academic Publishers, Dordrecht, 299 pp

Thompson R, Oldfield F (1986) Environmental magnetism. Allen and Unwin, London, 227 pp

Articles

Beck M Jr, Burmester R, Cembrano J, Drake R, Garcia A, Hervé F, Munizaga F (2000). Paleomagnetism of the North Patagonia batholith, southern Chile. An exercise in shape analysis. Tectonophysics 326:185–202

Besse J, Courtillot V (1991) Revised and synthetic apparent polar wander paths of the African, Eurasian, North American and Indian plates, and true polar wander since 200 Ma. J Geophys Res 96:4029–4050

Briden JC, Irving E (1964) Paleoclimatic spectra of sedimentary paleoclimatic indicators. In: Nairn AEM (ed) Problems in paleoclimatology. Wiley Interscience, New York, pp 199–250

Bullard EC, Everett JE, Smith AG (1965) The fit of the continents around the Atlantic. Philos T Roy Soc A 258:41–51

Crouzet C, Rochette P, Ménard G (2001) Experimental evaluation of thermal recording of successive polarities during uplift of metasediments. Geophys J Int 145:771–785

Galdéano A, Ciminale M (1987) Aeromagnetic evidence of the rotation of Sardinia (Mediterranean Sea): Comparison with the paleomagnetic measurements. Earth Planet Sc Lett 82:193–205

Gandino A, Lanza R, La Torre P, Manzoni M (1989) The role of remanent magnetization in magnetic interpretation of the Cimini-Vico area. Boll Geofis Teor Appl XXXI:233–244

Irving E, Major A (1964) Post-depositional detrital remanent magnetization in a synthetic sediment. Sedimentology 3:135–143

Opdyke ND (1972) Paleomagnetism of deep-sea cores. Rev Geophys Space Physics 10:213–249

Rolph TC (1997) An investigation of the magnetic variation within two recent lava flows. Geophys J Int 130:125–136

Valet JP, Meynadier L (1993) Geomagnetic field intensity and reversals during the past four million years. Nature 366:234–238

Van der Voo R (1990) Phanerozoic paleomagnetic poles from Europe and North America and comparison with continental rescontructions. Rev Geophys 28:167–206

Verosub KL (1977) Depositional and postdepositional processes in the magnetization of sediments. Rev Geophys Space Physics 15:129–143

Chapter 5

Magnetic Fabric of Rocks

The ordered structure of elementary particles in a crystal means that the magnetic properties of a mineral depend on the direction of the magnetic field relative to its crystallographic axes. Hence crystals are intrinsically magnetically anisotropic. In addition, some minerals have a direction of easy magnetization that depends on the shape of their grains (Chap. 2). Consequently, the magnetic anisotropy of a rock depends on the anisotropy of the individual grains of its minerals and their spatial arrangement. If the grains, even if highly anisotropic, are distributed randomly within a rock, the rock will have little or no anisotropy; if they are arranged according to preferential directions, then the rock as a whole will be anisotropic. The main geological application of magnetic anisotropy measurements is thus the study of the petrofabric. Knowing the rock minerals and their magnetic anisotropy characteristics, it is possible to determine the spatial distribution of the grains, which results from the various forces acting when the rock was formed. With respect to other methods of petrofabric analysis, magnetic anisotropy has some big advantages: it has a very high resolution power, it is quick, cost-effective and above all it can be performed systematically on all outcrops. For example, the determination of paleocurrents on the grounds of field observation requires finding mesoscopic structures such as flute casts and foreset laminae, which are not always present or easy to find out; on the other hand, the grain alignment in any sample taken from any outcrop can easily be determined magnetically.

5.1
Magnetic Anisotropy

All magnetic properties vary according to direction and therefore there are various types of anisotropy. Broadly speaking, a distinction can be made between anisotropy of magnetic susceptibility (AMS) and anisotropy of remanence (AIRM, anisotropy of isothermal remanent magnetization; AARM, anisotropy of anhysteretic remanent magnetization; etc.). We will mainly examine AMS, the one that so far has found the most applications in geology.

The basic relationship between induced magnetization and the external applied field is linear for diamagnetic minerals

$$J_i = \kappa H \tag{5.1}$$

and, providing H is low, may be assumed as linear for paramagnetic and ferromagnetic minerals also, as shown by the trend of the Langevin function (Fig. 2.4) and by the hysteresis cycle (Fig. 2.14), respectively. Since the Earth's field is $F < 70\ \mu T$ ($H_E < 55\ A\ m^{-1}$), the approximation of linearity is valid in geological studies and we can limit ourselves to the low-field AMS. In the case of an isotropic substance, magnetic susceptibility is represented by a single constant, κ (Eq. 5.1); if the substance is anisotropic, it is represented by a set of constants (k_{ij}) that form a second-rank symmetric tensor. The relationship between J_i and H is expressed by the equations (writing J instead of J_i from now on)

$$J_1 = k_{11}H_1 + k_{12}H_2 + k_{13}H_3$$

$$J_2 = k_{21}H_1 + k_{22}H_2 + k_{23}H_3 \tag{5.2}$$

$$J_3 = k_{31}H_1 + k_{32}H_2 + k_{33}H_3$$

with $(1,2,3)$ being a Cartesian coordinate system and $k_{ij} = k_{ji}$. Among all possible Cartesian reference systems, there is one for which the non-diagonal terms of the tensor cancel each other out so that Eqs. 5.2 simplify to

$$J_1 = k_{11}H_1$$

$$J_2 = k_{22}H_2 \tag{5.3}$$

$$J_3 = k_{33}H_3$$

The three values k_{11}, k_{22}, k_{33} are the eigenvalues of the tensor: they are called the principal susceptibilities and are normally indicated with the symbols k_{max}, k_{int}, k_{min} (or $k_1 > k_2 > k_3$) with the corresponding directions being the principal directions (maximum, intermediate, minimum). This tensor is represented by a tri-axial ellipsoid (Fig. 5.1): the axes coincide with the principal directions and their lengths are proportional to the eigenvalues. The anisotropy of a specimen can be represented visually by the shape of the ellipsoid; the eigenvalues are usually normalized based on the mean value k_m, corresponding to the bulk susceptibility κ.

The values k_1, k_2, k_3 can be combined in various manners to calculate parameters which describe the shape of the ellipsoid; the simplest and most widely used are

$$P = k_1 / k_3; \quad L = k_1 / k_2; \quad F = k_2 / k_3$$

$$P_J = \exp\sqrt{\{2\ [(\eta_1 - \eta)^2 + (\eta_2 - \eta)^2 + (\eta_3 - \eta)^2]\}} \tag{5.4}$$

$$T = 2\ln(k_2/k_3) / \ln(k_1/k_3) - 1 = (\ln F - \ln L) / (\ln F + \ln L)$$

where $\eta_i = \ln k_i$ and $\eta = (\eta_1 + \eta_2 + \eta_3) / 3$.

The degree of anisotropy P is a measure of how marked anisotropy is. In most minerals, $1 < P < 1.7$, although in some cases it can be $P > 100$; in rocks it is almost always $P < 1.3$–1.4, and values as low as $P \approx 1.005$ are still significantly measurable, highlighting barely hinted preferential alignments in the order of 1%. The corrected (or Jelinek) anisotropy degree P_J is preferred in most cases, as it incorporates the three values and

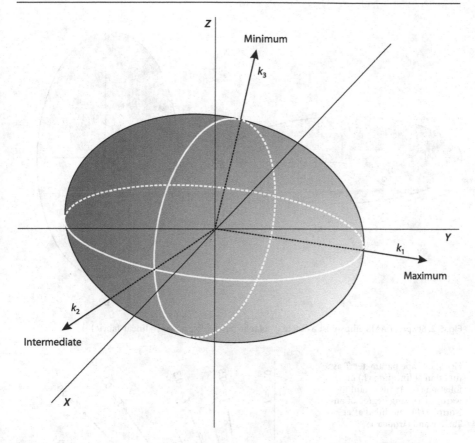

Fig. 5.1. Anisotropy of magnetic susceptibility is geometrically represented by a tri-axial ellipsoid with axes $k_1 > k_2 > k_3$

thus gives the more complete information. The magnetic lineation and foliation (L, F) describe the shape of the ellipsoid. Referring, for the sake of simplicity, to revolution ellipsoids (Fig. 5.2), L prevails when the ellipsoid is prolate $(k_1 > k_2 = k_3)$, and F when it is oblate $(k_1 = k_2 > k_3)$. A more detailed evaluation of the shape of the ellipsoid is given by the Jelinek shape parameter, T, for which $-1 \leq T < 0$ corresponds to prolate ellipsoids, and $0 < T \leq 1$ to oblate ellipsoids. Figure 5.3 is similar to that of Flinn's, used in structural geology, and it shows the relationships between L, F and T. The terms magnetic lineation and foliation also have a geometrical meaning. The lineation corresponds to the direction of k_1, the foliation to the plane being defined by the directions of k_1 and k_2 and hence orthogonal to k_3.

The AMS directional data are usually represented in stereographic projection; dealing with axes, the projection is on the lower hemisphere, as in structural geology. It has become conventional to represent the principal directions k_1, k_2, k_3 by squares, triangles and dots, respectively. Considering a group of specimens from the same outcrop, if the rock fabric is foliated, the k_3-axes are grouped, while k_1 and k_2 are more or

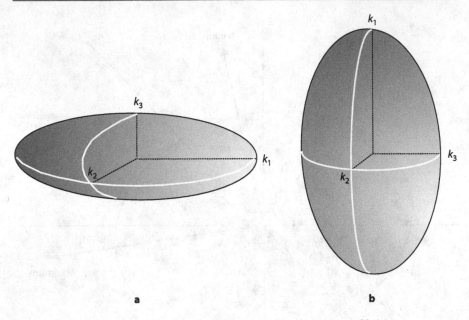

Fig. 5.2. Shape of AMS ellipsoids; **a** oblate (= planar fabric); **b** prolate (= linear fabric)

Fig. 5.3. Shape parameter T as a function of lineation (L) and foliation (F). *Arrows* point towards increasing degree of anisotropy (P) (modified after Tarling and Hrouda 1993)

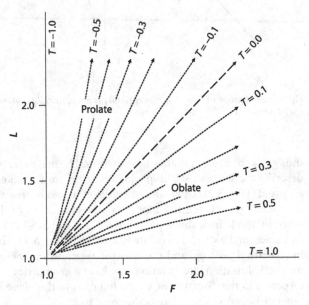

less dispersed along a girdle within the foliation plane (Fig. 5.4a); if lineation prevails, the axes k_1 are grouped (Fig. 5.4b) and there is no coherent foliation; if both lineation and foliation are developed, then each of the three directions forms a fairly well defined cluster (Fig. 5.4c). The statistical analysis of the AMS data is made complex by

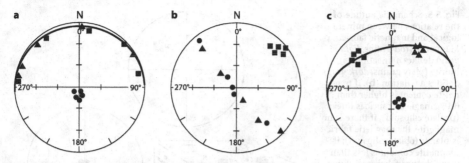

Fig. 5.4. Equal-area projection of the principal susceptibility directions in various types of magnetic fabric. Symbols: *square* = k_1; *triangle* = k_2; *dot* = k_3; *great circle* = magnetic foliation; **a** well-grouped k_3-axes, foliation well developed: the fabric is planar; **b** well-grouped k_1-axes, lineation well developed: the fabric is linear; **c** all axes well-grouped, both foliation and lineation developed

the fact that susceptibility is a tensor: the principal directions are not vectors and their distribution is not Fisherian. Various methods have been proposed: the simplest one calculates the terms of the mean tensor K_{ij} as the means of the corresponding N terms k_{ij} of the individual normalized tensors

$$K_{ij} = \sum(k_{ij}) / N \tag{5.5}$$

The dispersion of the mean principal directions is represented in stereographic projection by confidence ellipses. However, if the AMS derives from the superposition of fabrics of different origin and orientation, then the distribution is not unimodal and the mean values are no longer significant; the contour diagram of the principal directions can highlight significant concentrations in semi-quantitative fashion.

This problem brings us to a more thorough examination of the meaning of AMS data. The NRM of a rock can consist of multiple components (Chap. 4), each carried by grain populations that differ in mineralogical composition, domain state, etc. One of the most important steps in any paleomagnetic study consists in identifying and separating the various components. The situation is similar, but more complex, in AMS studies. While the NRM is carried by ferromagnetic minerals only, all minerals contribute to the AMS of a rock. The AMS of ferromagnetic minerals is usually distinguished from that of the para- and diamagnetic matrix. If the percentage of ferromagnetic minerals exceeds 0.1% ($\kappa > 3\,000$–$4\,000$ μSI), then the measured low-field AMS is not substantially influenced by the matrix (Fig. 5.5), but this can become dominant when $\kappa < 500$ μSI. For intermediate values, it may be necessary to separate the two contributions, which is possible with high-field AMS measurements. High fields saturate the signal carried by ferromagnetic minerals, which thus becomes distinguishable from that of dia- and paramagnetic minerals, which instead continues to change as a function of H.

Magnetic fabric analysis is used mostly in the case of rocks containing magnetite, hematite, biotite or chlorite. Hematite is, together with pyrrhotite, the most anisotropic mineral, as $P > 100$. Their anisotropy is magneto-crystalline, due to interaction between the reticular forces and the spin of the electrons: the direction of easy mag-

Fig. 5.5. Schematic outline of the relations between mineralogical and magnetic fabric. Magnetite (rod grains, *light gray*) defines a lineation (**a**), biotite (platy grains, *dark gray*) defines a foliation (**b**). If magnetite content is higher than 0.1%, magnetic fabric is linear (prolate ellipsoid); if there is no magnetite, the magnetic fabric is planar (oblate ellipsoid); if magnetite content is less than 0.1%, the two fabrics are superimposed one upon another

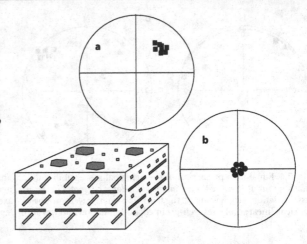

netization (k_1) lies in the basal plane and the hard direction (k_3) is orthogonal to it. Similarly, in the case of biotite and most phyllosilicates, the directions k_1 and k_3 are respectively parallel and orthogonal to the cleavage plane, but the degree of anisotropy is much smaller, $P < 1.35$ for biotite and $1.2 < P < 1.7$ for chlorite. The susceptibility ellipsoid of all these minerals is typically oblate. Magnetite has mainly shape anisotropy, whose characteristics vary according to whether the grains are single- (SD) or multi-domain (MD). The susceptibility of ferromagnetic minerals depends on the fact that the external magnetic field modifies the arrangement of the domain walls. If the grain is MD, the walls have a certain freedom of movement. The direction of easy magnetization is that of the elongation of the grain, so the AMS mimics the shape of the grain: k_1 is parallel to the long direction, k_3 is orthogonal to it. In the case of SD grain, the situation is reversed. Spontaneous magnetization is already in the long direction, corresponding to the minimum of magnetostatic energy, and hence an external field cannot change it in any way: the susceptibility measured parallel to the long dimension of the grain is necessarily nil (k_3), while in an orthogonal direction (k_1) the external field can induce a partial rotation of spontaneous magnetization. In both cases MD and SD, the susceptibility ellipsoid is prolate, but the relationship between the axis k_1 and the long direction of the grain is different: the two directions are parallel in the case MD, orthogonal in the case SD.

Although entirely SD magnetite is not common in rocks, whenever its presence is suspected the interpretation of the AMS data in terms of fabric must be accompanied by more thorough analyses, such as the study of one type of the remanence anisotropy, such as AIRM or AARM. Remanence anisotropy is caused mainly by the fact that the internal demagnetizing field of a grain is smallest in the long direction, greatest in the transverse direction, regardless of the domain type. Remanence intensity J_r is therefore greatest (r_1) when the field that caused it was parallel to the long direction of the grain, smallest (r_3) when it was orthogonal. Even if the relationship between J_r and H is not linear, as a first approximation the AIRM can be treated like AMS. Attributing then to $r_1 > r_2 > r_3$ the same meaning as $k_1 > k_2 > k_3$, in MD magnetite the shape of the two anisotropy ellipsoids is similar and the two sets of principal directions mutually coincide. In the SD case the shape is different and the largest and smallest axes exchange their position. Figure 5.6 schematically summarizes the results of measurements on a rock with elongated grains (rods) of magnetite. In the case of AMS, the projection in Fig. 5.6a corresponds to a planar fabric of SD grains, linear fabric of MD grains. Vice versa, the projection in Fig. 5.6b corresponds to a planar fabric for MD, linear for SD. The ambiguity is solved in the case of the AIRM, in which foliation (Fig. 5.6c) and lineation (Fig. 5.6d) do not depend on the state of the grain.

The anisotropy of most rocks is mainly due to that of the grains (magneto-crystalline or shape). Even isotropic grains, as in the case of idiomorphic magnetite, can cause an AMS, called distribution anisotropy, if they are close enough to interact with each other. In the presence of an external field H, a grain acquires a magnetic moment m, which in turn produces a microscopic induced field h. The microscopic field decreases as a function of the distance r from the grain, according to the relationship $h \propto m / r^3$, and its effects are felt over distances of the same order of magnitude as the dimensions of the grain. If the grains are all arranged in a line, then each individual grain is subjected to the external field H and to the microscopic field of the adjacent grains (Fig. 5.7). The microscopic field is added to the external one when H is parallel to the grain alignment, and subtracted when it is orthogonal. Susceptibility is therefore maximum when measured parallel to the alignment, smallest when measured transverse to it.

5.2
Laboratory Techniques

The procedures for collecting and orienting the samples are wholly similar to those used in paleomagnetism. Particular care must be taken in cutting the specimens: the shape of cubes and cylinders (with height/diameter ≈ 0.9) must be as regular as possible, to avoid introducing a shape anisotropy. When the same group of specimens is studied for both fabric and paleomagnetism, it is advisable to perform AMS measurements before demagnetization, which can introduce irreversible effects on the state of the ferromagnetic domains and, in the case of thermal demagnetization, mineralogical transformations.

The measure of susceptibility is based on the relationship $J = \kappa H$. The magnetic field can be steady or alternating, its intensity low or high. The most common measurement is the one called low-field susceptibility, which is performed in an alternating field with peak values in the order of a few hundreds of μT, slightly greater than the Earth's field. From Eq. 2.1, the following is obtained

$$B = \mu_0 H + \mu_0 J = \mu_0(H + \kappa H) = \mu_0(1 + \kappa)H = \mu H \qquad (5.6)$$

where μ is the magnetic permeability of the medium, usually expressed as $\mu = \mu_0 \mu_r$, with $\mu_r = 1 + \kappa$ being relative magnetic permeability. This last relationship reminds us that both μ_r and κ are numbers, i.e. dimensionless quantities. The characteristics of an electrical circuit containing inductive elements and carrying an alternating current depend on B, and hence on μ_r. For example, the inductance L of a solenoid coil with length l and formed by N turns of section A, inside a medium with relative permeability μ_r is

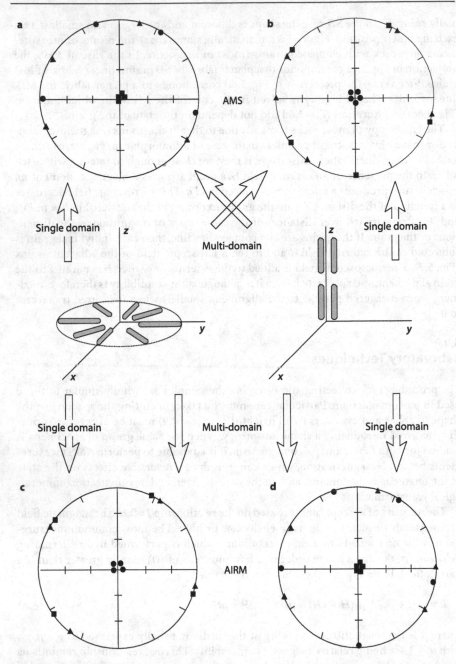

Fig. 5.6. Magnetic fabric in rocks with single-domain (SD) or multi-domain (MD) magnetite acicular grains. Symbols: *square* = k_1; *triangle* = k_2; *dot* = k_3. Anisotropy of magnetic susceptibility (AMS): cluster of k_1-axes (**a**) corresponds to either a linear fabric (MD grains) or a planar fabric (SD grains), cluster of k_3-axes (**b**) corresponds to either a planar fabric (MD) or a linear fabric (SD). Anisotropy of isothermal remanent magnetization (AIRM): for both SD and MD grains cluster of k_3-axes (**c**) corresponds to a planar fabric, cluster of k_1-axes (**d**) corresponds to a linear fabric (modified after Tarling and Hrouda 1993)

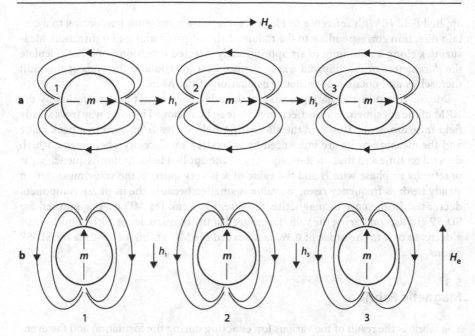

Fig. 5.7. Schematic picture of distribution anisotropy. An external field H magnetizes isotropic grains lined up at a distance similar to their diameter: each grain acquires a magnetic moment m and a microscopic induced field h, which acts on the neighboring grains. The total field H acting on grain 2 is $H_{tot} = H + h_1 + h_3$ if H is parallel to the alignment (**a**), $H_{tot} = H - h_1 - h_3$ if H is orthogonal (**b**) (modified after Stephenson 1994)

$$L = \mu_0 \mu_r N^2 A / l \tag{5.7}$$

A widely used instrument for measuring κ is the susceptibility bridge, which consists of two identical air-cored coils, at whose ends the measured signal is zero because $L_1 = L_2$. Inserting a rock specimen into a coil, the inductance connected thereto changes and therefore

$$\Delta L = L_1 - L_2 = \mu_0 N^2 A / l \cdot (\mu_{rs} - \mu_{ra}) \approx C(\kappa_s - \kappa_a) \tag{5.8}$$

where C is a constant of the instrument, μ_{rs}, κ_s the relative magnetic permeability and susceptibility of the specimen, μ_{ra}, κ_a those of air. As a first approximation, since $\kappa_a = 0.35\ \mu SI$, one has $\Delta L \propto \kappa_s$. Equation 5.8 presupposes that the coils are completely surrounded by the medium, either air or rock, but in fact the rock specimen is simply introduced into the coil. The appropriate corrections must therefore be made and, in any case, the signal must be normalized according to the volume of the specimen. The susceptibility κ is also called volume susceptibility; in various applications, mass susceptibility is of interest: it results from the relationship $\chi = \kappa / \varrho$, where ϱ is density, and its dimensions are $m^3\,kg^{-1}$.

To study AMS it is necessary to know the coefficients k_{ij} of the tensor in Eq. 5.2, then to measure susceptibility in different positions, turning the sample relative to the applied field H. With reference to Fig. 5.1, each value is measured according to a certain direction corresponding to the radius of the ellipsoid along that direction. Measuring κ along a minimum of six appropriately selected directions, one can calculate the parameters of the ellipsoid whose surface best interpolates the ends of the radii themselves and obtain a quantitative evaluation of the AMS.

Measurement in alternating fields has the advantage of not being affected by the NRM of the specimen, whose effect must instead be removed in some way from steady field measurements. However, the alternating field varies from zero to a peak value and the measurements are influenced by relaxation or viscosity phenomena, which depend on time and thus on the frequency of the applied field. At low frequencies, J is practically in phase with H and the value of κ is very close to the value measured in steady field; as frequency rises, the value is smaller, because the in-phase component decreases. In the case of magnetite, the effect is weak for MD grains, marked for SD-SP grains: comparing the values measured at the frequencies of 1 kHz and 10 kHz, a decrease of κ in the order of 0.3% is observed for MD grains, of 20–25% for SD-SP grains.

5.3
Magnetic Fabric

The fabric is the result of the various forces acting during the formation and the eventual geological history of the rock: mainly gravity, Earth's magnetic field, hydrodynamic forces and tectonic stress. All of these act according to their own direction and tend to orient crystals and grains, based on shape and size, according to preferential directions that correspond to the balance between the forces acting in each case. Considering for a moment these initial sentences, the extreme complexity of the problem is immediately apparent. The case of folded, detrital sedimentary rocks is an eloquent example: the fabric described by the AMS is the final result of deposition, diagenesis, burying and folding. The interpretation of the fabric can only use very schematic general models, which must be analyzed critically every time they are applied to a particular case.

5.3.1
Sedimentary Rocks

The forces that control grain deposition are the gravity, the current velocity, the Earth's magnetic field (for smaller ferromagnetic grains, Fig. 4.4), and, in the case of very dense suspension, the traction the grains exert on each other. Let us schematically classify the shape of the grains as flattened or elongated, respectively platy or rod. We can envision several scenarios. If sedimentation occurs in still water, in the absence of any current as in a lagoon or lake environment, the main force is gravity: platy grains will then deposit parallel to the bedding plane and the rod ones do the same with their greatest axis distributed randomly in that plane. The susceptibility ellipsoid is strongly oblate, with the axis k_3 orthogonal to the bedding (Fig. 5.8), well developed magnetic foliation and poorly defined or undefined lineation. The effect of a current is dual: platy grains tend to imbricate, so foliation no longer coincides with the bedding and plunges upstream; rod-shaped grains tend to align with their greatest axis parallel to

Fig. 5.8. Magnetic fabric of hematite-bearing clayey silts deposited in still water (Plio-Pleistocene continental sediments, NW Italy). Symbols: *square* = k_1; *dot* = k_3. Platy grains lie within the bedding plane: the k_3-axes are vertical, the k_1- and k_2-axes (not shown) are dispersed within the bedding plane, which coincides with the magnetic foliation

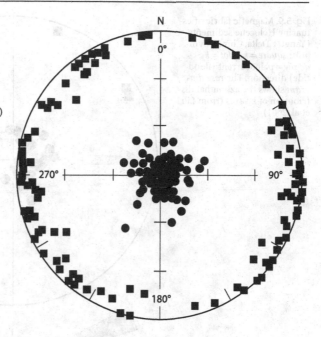

the flow. Imbrication of the magnetic foliation thus gives the absolute paleocurrent direction (Fig. 5.9); individual specimens lineation form a girdle and often crowd close to the flow direction. If deposition occurs on a slopy surface, rod grains can also roll, arranging themselves with their greatest axes orthogonal to the plunge of the bedding and, in the absence of current, a magnetic lineation is developed in this direction In the presence of a current, the distribution of lineation is the compromise between bedding slope and flow velocity.

Deposition is only the first stage in the formation of the rock; the grains can subsequently be reoriented as a result of the circulation or ejection of fluids, the formation of new minerals, reworking (bioturbation, slumping), compaction, etc. The final fabric is almost always characterized by a foliation that coincides with or approaches to (<20–25°) the bedding and the details need to be interpreted case by case, with the aid of any other available information.

An example of the great variability that may occur in the same type of sediment is shown schematically in Fig. 5.10, which refers to tephra levels interbedded in carbonate shelf deposits. Deposition took place at ambient temperature by settling of grains with average dimensions of 0.2–0.02 mm; the position of the vent is unknown, but certainly far from the shelf and the neighboring areas, where there is no other evidence of volcanic rocks. Far from the shelf, the deposition characteristics of the tephra levels are still the primary ones: the magnetic fabric is poorly developed and only a weak lineation occurs, nearly parallel to the direction of NRM. This would mean that the cooling and magnetization of magnetite grains occurred during wind transport: in the absence of other constraints, magnetization was acquired according to the easy direction, i.e. that of elongation. Depositions in still water (Fig. 5.10a) were driven by gravity and controlled by the Earth's magnetic field which, orienting the magnetic

Fig. 5.9. Magnetic fabric of estuarine Holocene sediments (Yangtze Delta, China). Symbols: *square* = k_1; *dot* = k_3; *arrow* = paleocurrent (flood-tide) direction. The *rose diagram* shows the azimuthal distribution of k_1-axes (from Liu et al. 2001)

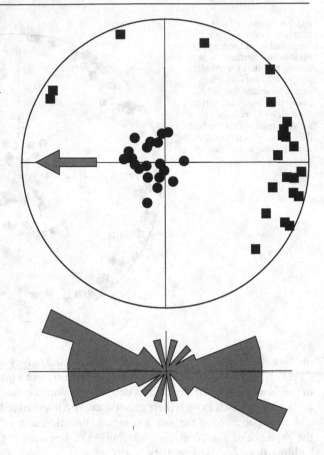

moment of the grains, would have oriented their long directions as well. However, in distal areas (Fig. 5.10b) there are also secondary deposits, with clear indication of slumping: the magnetic fabric is characterized by marked foliation close to the bedding and lineation orthogonal to the foliation plunge. In shallow water, proximal shelf areas (Fig. 5.10c) the tephra deposit was reworked by the action of storm waves and fabric is incoherent from specimen to specimen.

5.3.2
Igneous Rocks

Effusive, hypabyssal, intrusive and pyroclastic rocks all have different emplacement modes, controlled by gravity, magmatic or pyroclastic flow, and syn-emplacement stress. Moreover, while grains in sedimentary rocks can very often be considered as passive, in the case of magmatic rocks the crystal forming process can be synchronous with emplacement. The action of the magnetic field is almost always nil, because the emplacement temperature is higher than the Curie point of ferromagnetic miner-

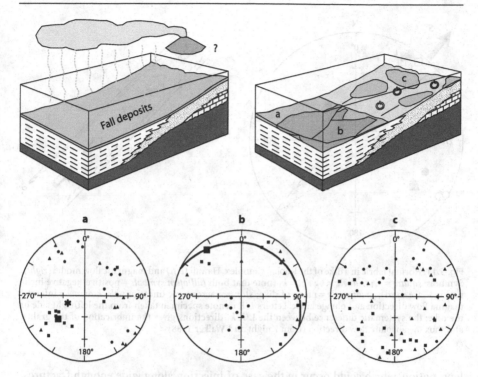

Fig. 5.10. Deposition of tephra in a carbonate shelf environment (*left*) and evolution of the deposit (*right*) (NW Apennines, Italy). Symbols: *square* = k_1; *triangle* = k_2; *dot* = k_3; *great circle* = magnetic foliation; *star* = site mean NRM direction (see text for further explanation). **a** The primary, depositional fabric is still preserved in distal deposits; **b** at the base of the shelf the fabric is typical of turbidites; **c** in shallow-water environment, the fabric is chaotic because of reworking by storm waves (modified after d'Atri et al. 1999)

als, except in the case of some pyroclastic rocks which can be deposited at temperatures of a few hundreds of degrees or even normal ambient temperatures.

In the case of lava flows, dikes and sills the starting model is similar to that of detrital rocks. Titanomagnetite grains are oriented with their greatest dimension parallel to the magma flow: magnetic lineation is close to the flow direction, foliation is parallel to the dike or sill walls, whereas in lava flows it matches the (paleo)topographic surface the lava flowed upon. In the case of basic dikes and sills, magnetite is usually present with contents >1% and controls the magnetic fabric. The model assumes a laminar motion parallel to the intruded fracture, which orients crystals according to the flow direction. In many cases, the model works and the fabric is said to be normal (Fig. 5.11): the axes k_3 and k_1 are respectively orthogonal and parallel to the plane of the dike. When mesoscopic indicators of the flow direction are found, a good correspondence with the direction of k_1 is observed (±25°). On the other hand, the systematic study of dike swarms (Fig. 5.12) has shown that the normal fabric is usually associated with an intermediate fabric (k_2 normal to the plane of the dike) and even a reverse fabric (k_1 normal to the plane). A random fabric may be an indication of turbu-

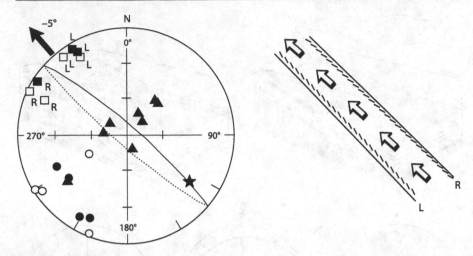

Fig. 5.11. Normal fabric in a dike of the Koolau Complex, Hawaii (*left*) and magmatic flow model (*right*). Symbols: *square* = k_1; *triangle* = k_2; *dot* = k_3 (note that both *full/open symbols* – positive/negative inclination – are used); *great circle* = plane of dike; *star* = macroscopic lineation; *black arrow* = inferred magma flow direction and plunge angle. Letters *L/R* denote specimen taken from the left/right side of the dike: the systematic difference between the L/R k_1 directions gives the imbrication of the grains, and thus the absolute flow direction (from Knight and Walker 1988)

lent motion, which could occur in the case of injection along wide enough fractures. On the one hand, these anomalous fabrics may result from stress simultaneous with the emplacement: contraction as a result of cooling, compaction caused by the lithostatic load of the overlying magma, regional tectonic stress. On the other hand, they are not necessarily anomalous: the reverse fabric may be caused by the presence of SD grains; the intermediate one may be caused by exchange of the two directions k_2, k_3, whose orientation relative to the plane of the dike, according to theoretical models, depends on the stress undergone by the flowing magma and by the elongation of the grain, i.e. the ratio between its minimum and maximum lengths.

The situation, already so complex for dikes, becomes difficult for lava flows, also because they are emplaced along the topographic paleosurface, which for a volcano may not be horizontal. The marked difference in cooling rate between the different parts of a flow complicates the situation even further, to the extreme case where consolidated lava blocks are mobilized again. Usually, the definition of foliation and lineation at the scale of the outcrop is less good than for dikes and the data of various outcrops of the same flow have a certain dispersion, especially when slope is gentle or the flow wanders across sub-horizontal surfaces. Close to a summit crater (Fig. 5.13) the slope is normally steeper and the lineation plunges radially downwards.

The emplacement modes of pyroclastic rocks are highly variable; at a large distance from the vent, tephra levels are substantially sedimentary rocks (see Sect. 5.3.1), whilst at a short distance ignimbrites and surges have such energy as to cause a chaotic motion of the transported solid particles, which – however – can assume orderly arrangements when their kinetic energy decreases and they deposit. The normal fabric of pyroclastic flows is considered the one with the foliation plunging upflow: the flow

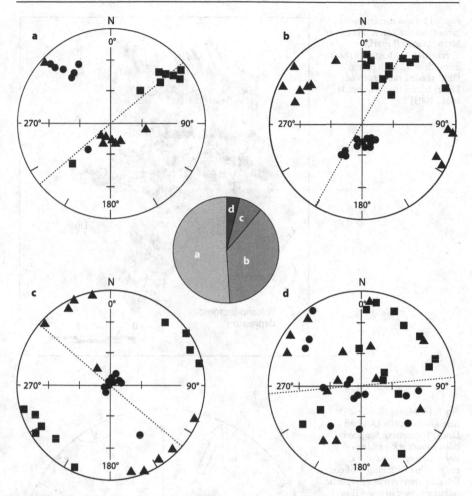

Fig. 5.12. Different types of fabric in 95 mafic dikes (Ponta Grossa dike swarm, Brasil) and pie diagram of their relative frequency. Symbols: *square* = k_1; *triangle* = k_2; *dot* = k_3; *dashed line* = dike plane tilted to vertical. Fabric type; **a** normal, **b** intermediate; **c** reverse; **d** chaotic (modified after Raposo and Ernesto 1994)

direction is given by the azimuth of the foliation pole (site mean k_3) and the lineation is close to it (Fig. 5.14). Systematic sampling allows tracing of the flow path back to the vent (Fig. 5.15). The flow directions at some sites may be discordant, as a consequence of local effects. The flow may follow the course of a paleovalley and hence locally have a very different direction, or climb over a slope or a crater rim: the solid material is first deposited and then flows back, rolling downwards by effect of gravity. When these disturbance effects are known, the measurements of AMS are almost always found to be reliable. For example, in the case of the Mount Vesuvius eruption of A.D. 79, which destroyed the Roman city of Pompeii, the pyroclastic flows that hit the city before it was completely buried followed the city walls and the streets (Plate 3):

Fig. 5.13. Flow directions of lavas from the summit crater of Mt. S. Angelo (Lipari, Tyrrhenian Sea). *Arrows* show azimuth of k_1, with plunge. DEM shaded relief map of Lipari Island from Gioncada et al. (2003)

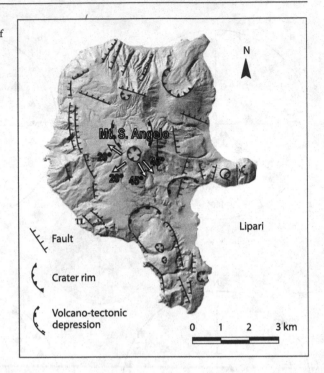

Fig. 5.14. Magnetic fabric in fine-grained tuffs (Aeolian Islands, Tyrrhenian Sea). Symbols: *square* = k_1; *triangle* = k_2; *dot* = k_3; *large symbol* = site mean value with ellipse of confidence; *great circle* = magnetic foliation; *open arrow* = flow direction. Magnetic foliation is tilted upstream and lineation plunges toward the vent

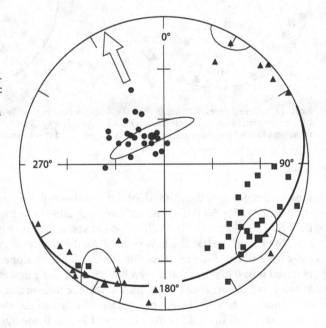

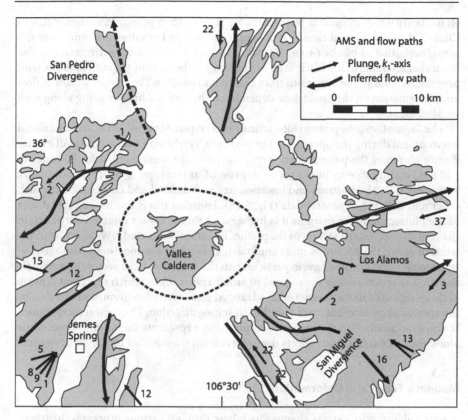

Fig. 5.15. Inferred flow paths in ash-flow tuffs (Valles Caldera, New Mexico, USA). *Small arrows* show azimuth of k_1-axis, (plunge positive downwards). Magnetic lineations converge towards the caldera; the diverted directions in the two divergence areas are regarded as a buttress effect of basement topographic highs (from MacDonald and Palmer 1990)

the direction of the flow is faithfully recorded by the magnetic fabric and the magnetic foliation coincides with the arrangement of the tiles from collapsed roofs incorporated in the deposit.

The acquisition of the fabric of effusive rocks may be represented by a very simple model: the flow generates the fabric, the rapid cooling freezes it, there is no tectonic stress. The emplacement of intrusive rocks is far more complex: it is simultaneous to magma cooling, it is correlated to a regional stress, it occurs over very long times. The fabric is therefore affected by the flow of the magma, the changes in its viscosity and the finite deformation it undergoes before complete crystallization. Moreover, if intrusion took place in the lower levels of the crust, the tectonic processes that led to the eventual exhumation may have superimposed a secondary deformational fabric on to the primary one. For this reason, the typical intrusive fabric is the one of granitic rocks, which intrude the upper levels of the crust. For them, it is essential first of all to clarify which mineral is responsible for the fabric: in principle, as mentioned in Chap. 2, the dominant fabric is that of magnetite in type I granites, also known as *ferromagnetic*, that of biotite (+ any ilmenite) in type S, or *paramagnetic* granites. The main characteristic of the fabric is its consistency at the scale of the entire pluton, with progressive, continuous variations from one area to another. The relative weight of flow and deformation on the final fabric depends on the tectonic history going along with emplacement.

The Monte Capanne pluton (Elba Island) was emplaced at 6–7 Ma in an extensional environment during the opening of the northern Tyrrhenian Sea. It is a small granodiorite pluton of the paramagnetic type, with average susceptibility in the order of 140 μSI, mainly due to biotite. The degree of anisotropy is low, almost always $1.01 < P < 1.03$. The foliation and lineation are well defined and consistent with each other and with the structural data (Fig. 5.16). Foliation dip is low in the central area of the pluton; along the margins it is higher and almost always directed outwards; in the central and eastern sectors of the pluton, lineation is directed NW-SE, while in the western sector the pattern is more confused. These data are consistent with those of traditional structural geology; in particular, in the central-eastern sector the magnetic lineation is systematically orthogonal to small aplitic dikes, which represent tension gashes originated in the magma at an advanced state of solidification, and can thus be interpreted as an excellent marker of the stretching direction. The central sector, which is topographically also the highest one, plausibly represents the feeding zone, while along the margin foliation reflects the deformation transmitted by the country rock.

5.3.3
Magnetic Fabric and Deformation

A rock subjected to a stress changes its fabric through various processes: brittle or ductile deformation, diffusion creep and recrystallization. This is the typical case of metamorphic rocks, but weak deformations can take place without the development of an actual metamorphism, e.g. even just in response to burial. Depending on the intensity of stress, on the time over which it is active and on physicochemical conditions, overprint will be more or less marked. As deformation increases, the primary fabric is modified more and more until it is completely obliterated. The interpretation of AMS results on deformed rocks can be done only in a broader context, including mineralogical analyses, traditional petrofabric techniques, structural geology surveys.

In the case of progressive deformation of sedimentary rocks, the general model is in good agreement with many experimental observations. Let us consider a rock with primary fabric subjected to a shortening parallel to the bedding: the axis parallel to the shortening becomes shorter, the normal one becomes longer. The initial magnetic fabric has the foliation coinciding with the bedding and k_3 orthogonal to it (*Stage a*, Fig. 5.17). When deformation is very weak, k_1 is arranged perpendicular to the shortening (*Stage b*, Fig. 5.17), while still remaining in the bedding. As deformation grows, k_3 is reoriented parallel to the shortening and k_2 normal to the bedding (*Stage c*, Fig. 5.17), until the sedimentary fabric is completely obliterated: k_1 is normal to the bedding, k_3 is in the direction of shortening and k_2 in the orthogonal direction (*Stage d*, Fig. 5.17). The shift from *Stage b*, to *c*, to *d* in Fig. 5.17 corresponds to a gradual defor-

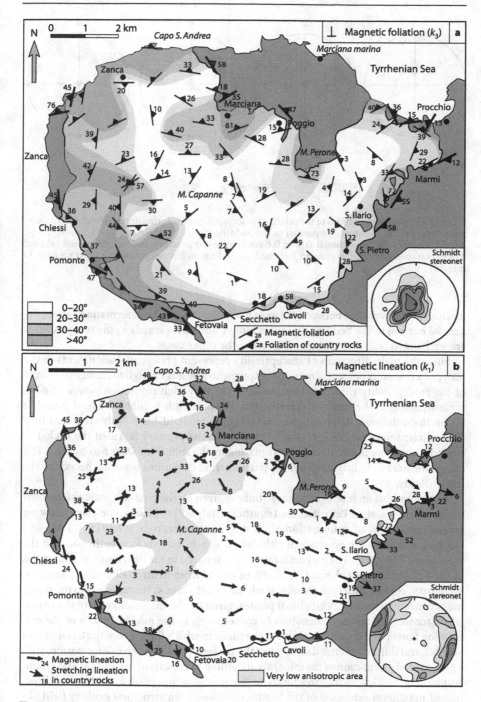

Fig. 5.16. Magnetic structural maps of Monte Capanne granodiorite pluton (Elba Island, northern Tyrrhenian Sea); **a** magnetic foliation and equal-area projection of k_3-axes; **b** magnetic lineation and equal-area projection of k_1-axes (granodiorite) and stretching lineation (country rock) (from Bouillin et al. 1993)

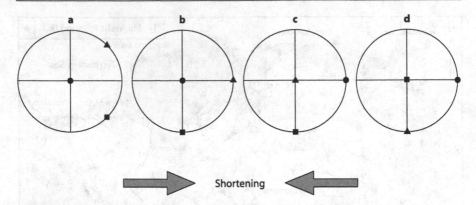

Fig. 5.17. Equal-area projections of the principal susceptibility axes during progressive deformation of sedimentary rocks (compression parallel to the bedding plane). Symbols: *square* = k_1; *triangle* = k_2; *dot* = k_3; **a** primary, depositional fabric; **b,c** superimposed primary and deformational fabrics; **d** deformational fabric: the maximum, intermediate and minimum susceptibility and strain axes do coincide

mation, while exchange between the axes is discontinuous. As deformation grows, the axis k_3, normal to the bedding, becomes longer and as it exceeds k_2 the two positions are exchanged with a 90° shift, and so on in the other cases.

Since the main directions of susceptibility represent preferential alignments in the order of a few units percent, a minimal response to stress is enough to change the shape of the susceptibility ellipsoid. This explains the very high resolution power of AMS measurements in the study of deformation. However, it is a doubled-edged sword: it is true that minimum deformations can be brought to light, but it is also true that they are sufficient to cause the loss of information on the primary fabric of the rock. For example, a very weak ductile deformation causing the shift from *Stage a* to *b* (Fig. 5.17), causes the axes k_1 to migrate from the paleocurrent directions towards the axis of the (micro)folds.

An example of an interdisciplinary study is given by the post-orogenic extensional basins of the Calabro-Peloritan arc (southern Italy). Plio-Pleistocene clays show no macroscopic signs of deformation; their degree of anisotropy is low ($P_J < 1.05$) and can be ascribed to the presence of chlorite, in which the axis k_3 is orthogonal to the basal plane (Sect. 5.1). The magnetic fabric is compared to the distribution of the poles of the basal planes of chlorite (Fig. 5.18), obtained from neutron diffraction analysis. Systematic correspondences are observed: the direction of k_3 corresponds to the maximum concentration of chlorite basal planes, given by the maximum axis of the orientation tensor; the magnetic lineation k_1 corresponds to the minimum axis of the orientation tensor and is perpendicular to the plane in which the poles of the basal planes lie. AMS and diffractometric data suggest the presence of a pervasive deformation, with a preferential orientation of the minerals according to the axis of microfolds and crenulations (Fig. 5.19), which corresponds to the magnetic lineation k_1 and to the direction of maximum extension of the basins, obtained from structural geology field observations.

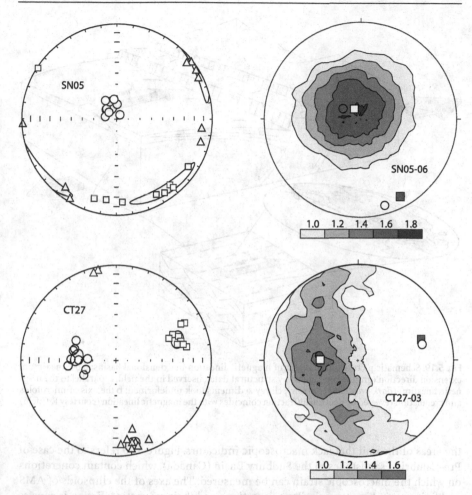

Fig. 5.18. Equal-area projection of the AMS (*left*) and neutron diffraction (*right*) data from clayey sediments from extensional basins in southern Italy. Symbols: AMS: *square* = k_1; *triangle* = k_2; *dot* = k_3 (*open symbols* used for sake of clarity); neutron diffraction: contour density of the poles of the chlorite basal planes; *square* = maximum axis of orientation tensor; *dot* = miminum axis of orientation tensor (courtesy F. Cifelli)

The fabric of the rock, once the primary one is obliterated (*Stage d*, Fig. 5.17), is purely deformational: the principal directions of susceptibility and of finite strain coincide and continue to coincide with increasing deformation. If the shapes of the two ellipsoids were also mutually correlated, the AMS could be used to quantify the strain, with the obvious advantage of being able to do so in all cases in which there are no macroscopic strain indicators. No general relationship was found and the correlations found in several particular cases could not be extrapolated to different situations. However, if a correlation does exist within a certain geological unit (sedimentary formation, intrusive massif, etc.) it can be used to extend the strain analysis to

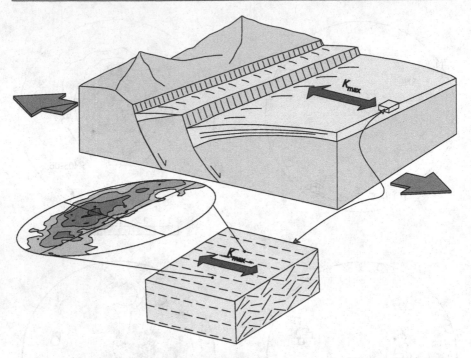

Fig. 5.19. Schematic picture of the origin of magnetic lineation in extensional basins; **a** the maximum extension direction (*gray arrows*), given by structural data observed in the field, is parallel to the magnetic lineation (*dark gray double-arrow*); clayey sediments look undeformed; **b** the axis of microfolds and crenulation revealed by neutron diffraction coincides with the magnetic lineation (courtesy F. Cifelli)

the areas of the unit that lack macroscopic indicators. Figure 5.20 refers to the case of Pre-Cambrian sandstones in the Sudbury Basin (Canada), which contain concretions on which the macroscopic strain can be measured. The axes of the ellipsoids of AMS and finite strain have very similar orientations and their magnetic foliation is consistent with the cleavage plane. Between the two degrees of anisotropy, $P_J(k)$, and of finite strain, $P_J(e)$ – defined by a formula similar to Eq. 5.4, in which the three main strains $e_1 > e_2 > e_3$ replace the three eigenvalues k_1, k_2 and k_3 – a good linear correlation exists (Fig. 5.21), significant to the 95% level.

5.4
Anisotropy and Remanence

At the end of a chapter dedicated to magnetic anisotropy, a question naturally emerges. Does the direction of the ChRM of a rock really correspond to that of the Earth's magnetic field F present when the rock is formed? As crystals have easy magnetization directions and are often arranged according to preferential directions, the doubt is well founded. The fabric may affect remanence and the example of inclination shallowing (Sect. 4.1.3) is eloquent: an orderly arrangement of the grains can introduce deviations of up to 10°–20°.

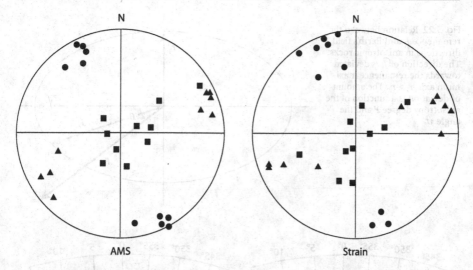

Fig. 5.20. Equal-area projection of the site mean principal directions of AMS (k) and finite strain (e) in the Chelmsford Formation (Sudbury Basin, Ontario, Canada). Symbols: *square* = k_1, e_1; *triangle* = k_2, e_2; *dot* = k_3, e_3 (from Hirt et al. 1988)

Fig. 5.21. Correlation of AMS, $P_j(k)$, versus finite strain, $P_j(e)$, in the Chelmsford Formation (Sudbury Basin, Ontario, Canada) (from Borradaile 1991)

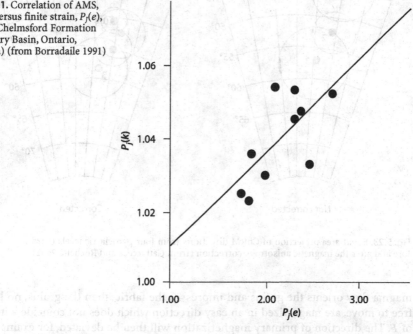

Inclination shallowing is a secondary effect: the grains, already magnetized in an easy direction, are oriented by the field F while they are deposited and are then deviated by the action of gravity. In igneous rocks, deviation can be primary: first the

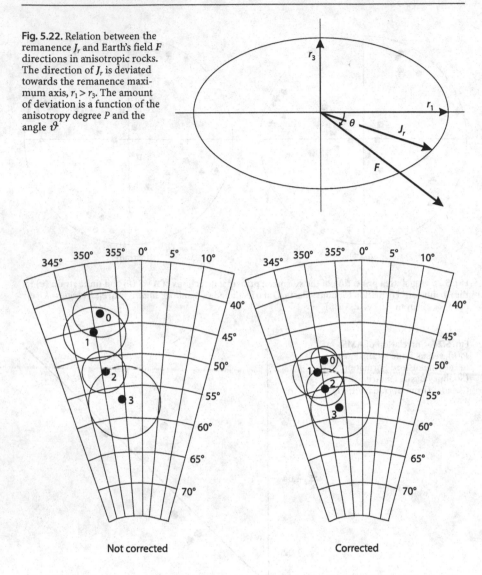

Fig. 5.22. Relation between the remanence J_r and Earth's field F directions in anisotropic rocks. The direction of J_r is deviated towards the remanence maximum axis, $r_1 > r_3$. The amount of deviation is a function of the anisotropy degree P and the angle ϑ

Not corrected

Corrected

Fig. 5.23. Equal-area projection of ChRM directions from four pyroclastic levels (Sardinia, Italy) before and after the magnetic anisotropy correction (from Gattacceca and Rochette 2002)

magma flow orients the grains and impresses the fabric, then the grains, no longer free to move, are magnetized in an easy direction which does not coincide with that of F. The direction of primary magnetization will then be deviated, for example approaching the magnetic foliation plane (Fig. 5.22). What saves paleomagnetism is the fact that on the one hand the rocks are not homogeneous: grains have different shapes, dimensions and magnetic properties. On the other hand, that preferential orientations are nearly always poorly developed, in other words that the degree of anisotropy is low. Anyway, it is clear that if the interpretation of the measurements wants fully to exploit the high precision of today's instruments, the presence of possible systematic errors must be controlled by measuring remanence anisotropy.

Figure 5.23 refers to four superposed ignimbrite levels that, based on field evidence, were emplaced during a very short time interval and therefore should have very similar directions of ChRM. Instead, they differ up to about fifteen degrees, which could be explained as a consequence of secular variation over a few kyr. AARM (anisotropy of anhysteretic remanent magnetization) measurements highlight a degree of anisotropy $1.04 < P_{ARM} < 1.31$. Correcting directions according to the values of P_{ARM}, dispersion is considerably reduced: the circles of confidence α_{95} overlap and the four directions become statistically undistinguishable.

In conclusion, the smaller the magnetic anisotropy of a rock the safer will be the assumption that its ChRM direction represents that of the past magnetic field. Vice versa, when the magnetic fabric is well developed, significant differences must be expected.

Suggested Readings and Sources of Figures

Books

Collinson DW (1983) Methods in rock magnetism and palaeomagnetism. Chapman and Hall, London, 503 pp
Dunlop DJ, Özdemir Ö (1997) Rock magnetism. Fundamentals and frontiers. Cambridge University Press, Cambridge, UK, 573 pp
O'Reilly W (1984) Rock and mineral magnetism. Blackie, Glasgow, UK, 220 pp
Stacey FD, Banerjee SK (1974) The physical principles of rock magnetism. Elsevier, Amsterdam, 195 pp
Tarling DH, Hrouda F (1993) The magnetic anisotropy of rocks. Chapman and Hall, London, 217 pp

Articles

Borradaile GJ (1991) Correlation of strain with anisotropy of magnetic susceptibility (AMS). Pure Appl Geophys 135:15–29
Bouillin J-P, Bouchez J-L, Lespinasse P, Pêcher A (1993) Granite emplacement in an extensional setting: An AMS study of the magmatic structures of Monte Capanne (Elba, Italy). Earth Planet Sc Lett 118:263–279
d'Atri A, Dela Pierre F, Lanza R, Ruffini R (1999) Distinguishing primary and resedimented vitric volcaniclastic layers in the Burdigalian carbonate shelf deposits in Monferrato (NW Italy). Sed Geol 129:143–163
Gattacceca J, Rochette P (2002) Pseudopaleosecular variation due to remanence anisotropy in a pyroclastic flow succession. Geophys Res Lett 29 (DOI 10.10129/2002GL014697)
Gioncada A, Mazzuoli R, Bisson M, Pareschi MT (2003) Petrology of volcanic products younger than 42 ka on the Lipari-Vulcano complex (Aeolian Islands, Italy): An example of volcanism controlled by tectonics. J Volcanol Geoth Res 122:191–220
Gurioli L, Pareschi MT, Zanella E, Lanza R, Deluca E, Bisson M (2005) Interaction of pyroclastic density currents with human settlements: Evidence fron ancient Pompeii. Geology 33:441–444 (DOI 10.1130/G21294.1)
Hirt AM, Lowrie W, Clendenen WS, Kligfield R (1988) The correlation of magnetic anisotropy with strain in the Chelmsford Formation of the Sudbury Basin, Ontario. Tectonophysics 145:177–189
Knight MD, Walker PL (1988) Magma flow directions in dikes of the Koolau Complex, Oahu, determined from magnetic fabric studies. J Geophys Res 93:4301–4319
Liu B, Saito Y, Yamazaki T, Abdeldayem A, Oda H, Hori K, Zhao Q (2001) Paleocurrent analysis for the late Pleistocene-Holocene incised-valley fill of the Yangtze delta, China by using anisotropy of magnetic suceptibility data. Mar Geol 176:175–189

MacDonald WD, Palmer HC (1990) Flow directions in ash-tuffs: A comparison of geological and magnetic susceptibility measurements, Tshirege member (upper Bandelier Tuff), Valles caldera, New Mexico, USA. Bull Volcanol 53:45–59

Raposo MIB, Ernesto M (1994) Anisotropy of magnetic susceptibility in the Ponta Grossa dyke swarm (Brazil) and its relationship with magma flow direction. Phys Earth Planet In 87:183–196

Stephenson A (1994) Distribution anisotropy: Two simple models for magnetic lineation and foliation. Phys Earth Planet In 82:49–53

Chapter 6

Magnetic Signature of the Earth's Crust

The Earth's magnetic field is generated in the core and in the lithosphere, as shown by the spherical harmonic analysis. In Chap. 1 we saw the characteristics of the core field and the geodynamo model; Chap. 3 discusses the magnetic anomalies caused by magnetization contrasts between geological structures of mining and regional tectonic interest. This chapter is dedicated to the magnetic characteristics of the Earth's crust as a whole: they are systematically different on continents and oceans and provide indications on large scale geodynamic processes. We will consider only the crust, because the magnetic role of the upper mantle is controversial and in any case secondary. Under normal conditions, ferromagnetism of rock minerals disappears at depths in the order of 30–60 km, given that:

- the Curie temperatures T_c of the various ferromagnetic minerals *s.l.* are lower than 670 °C, that of Fe is $\approx$750 °C;
- T_c variation as a function of pressure seems very limited; laboratory analyses on magnetite yield +2 °C kbar^{-1}, and hence an hypothetical T_c increase less than 10% for pressures in the order of 20 kbar;
- the temperature rise with increasing depth, the so-called geothermal gradient, has a mean value of 20 °C km^{-1}. The Curie isotherm, i.e. the surface below which ferromagnetic minerals lose their properties and behave as paramagnetic, is therefore located at a depth of some 30–35 km; it approaches the Earth's surface in regions with intense volcanic activity, and it deepens below cratonic regions where thermal flux is lowest.

The study of crustal anomalies requires a logistical support suitable to large scale surveys and it has developed over two parallel routes: satellite surveys, performed from altitudes of a few hundred kilometers, and shipborne and airborne surveys, performed along or near the Earth's surface. Satellite surveys (Sect. 3.4) analyze mostly the anomalies with wavelengths in the order of a thousand kilometers, due to the general characteristics of the crust or to large structures of at least sub-continental scale. Shipborne and airborne surveys investigate anomalies with shorter wavelengths, up to a few tens or hundreds of kilometers, and are mainly aimed at identifying the lateral magnetization contrasts connected to regional tectonic structures.

The different structure, composition and geodynamic history of the oceanic and the continental crust also reflect in the magnetic anomalies they produce. Oceanic anomalies (Fig. 6.1) have a regular general pattern, formed by alternating positive and negative bands, whose width is in the order of tens of kilometers and whose length can reach 1 000 km. Hence the *linear magnetic anomaly* term. Above continents, in-

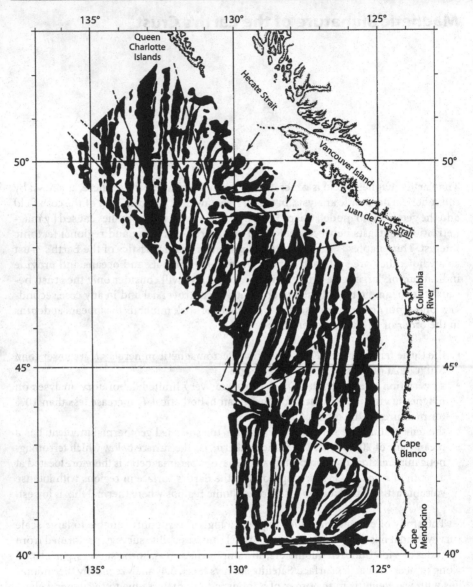

Fig. 6.1. Total magnetic field anomaly map southwest of Vancouver Island. Symbols: *black/white areas* = positive/negative anomaly; *straight lines* = faults offsetting the anomaly pattern (from Raff and Mason 1961)

stead, anomalies have a marked individual nature: they can be correlated to individual geological bodies or they can be arranged according to large scale trends that underline larger structures (Plate 2), while not being directly linked to the details of their geometry.

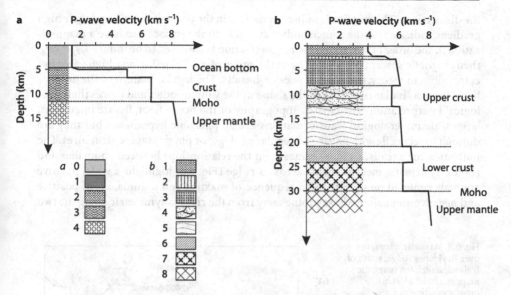

Fig. 6.2. Seismic structure of the Earth's crust; **a** oceanic crust according to Raitt (1963) and lithological model: *0* = sea water; *1* = sediments; *2* = basalt; *3* = gabbro; *4* = peridotite; **b** continental crust and lithologies in hercynian Europe according to Müller (1977): *1* = Cenozoic sediments; *2* = Mesozoic and Palaeozoic sediments; *3* = low-grade metamorphics; *4* = granite; *5* = migmatites; *6* = amphibolite; *7* = granulite; *8* = ultramafics

The structure of the two types of crust has been deduced on the basis of variations in the velocity of seismic waves. According to the Raitt model (Fig. 6.2a) the oceanic crust, on average 7 km thick, is formed by three layers. Layer 1 consists of sediments, Layer 2 of basaltic rocks, Layer 3, according to current interpretations, of gabbros. The continental crust, with a mean thickness of 30–35 km, has a far more complex structure; assuming as a reference the model proposed by Müller for the hercynian crust of central Europe (Fig. 6.2b), it is subdivided into upper (sediments, low grade metamorphic rocks with granitic intrusions, granitic laccoliths, migmatites) and lower crust (amphibolites, granulites). The role of granulites, eclogites and lherzolites in the crust-mantle transition region, the Moho, is still the subject of speculation. Lastly, we recall that the age of the oldest oceanic crust is less than 200 Ma (middle Jurassic), whereas the age of the cratonic continental crust is greater than 500 Ma and may reach 4 Ga (Archean).

6.1
Oceanic Crust

The pattern of oceanic anomalies, in addition to being highly regular, is also dislocated in correspondence of the main fracture zones of the ocean floor (Fig. 6.1), with an offset of even hundreds of kilometers. This demonstrates, on the one hand, the rigidity of the floor, on the other hand that anomalous structures were formed before the dislocations. The amplitude of the anomalies, in the order of 500 nT, and their high gradient, indicate that the source bodies are close to the surface and have a magnetization in the order of 5 A m^{-1}. If the magnetization is assumed to be induced, $J_i = \kappa H_E$, then magnetic susceptibility must be in the order of $\kappa \approx 10^5 \, \mu SI$, a very high value but compatible, in theory, with that of oceanic basalts. The regular pattern of the anomalies reflects a similar regularity in the shape of the source bodies and makes their geological interpretation problematic. Topography of the ocean floor, fissure intrusions, large scale mineralogical transformations are all plausible hypotheses, but they are difficult to reconcile with linear and regular geologic or physiographic structures. The indication for a realistic model comes from the relationships between anomalies and ridges. Systematic measurements across a ridge (Fig. 6.3) highlight a wide, positive anomaly centered on the rift and a sequence of maxima and minima, i.e. of positive and negative anomalies, whose profile away from the ridge is symmetric along the two

Fig. 6.3. Magnetic anomalies over Reykjanes Ridge, south of Iceland; **a** skeleton magnetic map; symbols: *black/white areas* = positive/negative anomaly; *A* = central anomaly over the ridge; **b** total field magnetic anomaly profiles projected perpendicular to the ridge axis (1 gamma = 1 nT) (from Heirtzler et al. 1966)

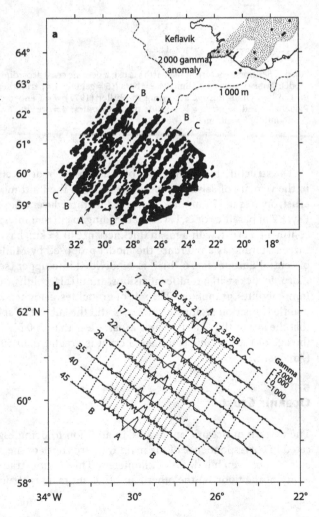

sides. Parallelism between the anomalies and the ridge axis, and symmetry of the maxima and minima sequence are the bases of the model proposed by Vine and Matthews and independently by Morley in 1963: they are the combined effect of the sea-floor expansion and the polarity reversal of the Earth's magnetic field. The model (Fig. 6.4) presupposes that new oceanic crust is formed in the rift region, where basaltic lavas are produced by the upward current of a mantle convection cell. The lavas

Fig. 6.4. The Vine-Matthews-Morley model: example for the Juan de Fuca Ridge (off the Pacific coast of Canada); **a** sea-floor spreading and magnetization: *black/white* = normal/reverse polarity magnetization in Layer 2; **b** magnetic anomaly map: *black/white* = positive/negative anomaly; **c** total field anomaly profile measured along the central portion of the map (1 gamma = 1 nT); **d** magnetization model and computed total field anomaly (from Vine 1968)

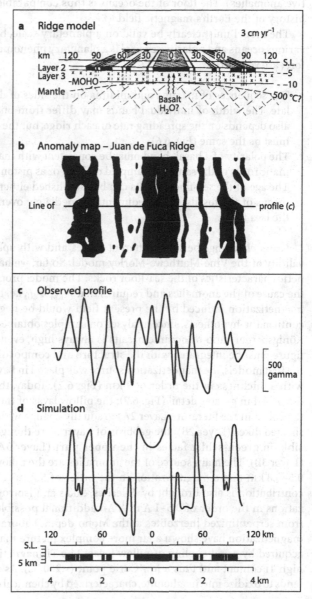

contain Ti-magnetite and as they cool below its Curie point they become magnetized in the direction of the Earth's field. As expansion proceeds, basalts move away from the rift, new lavas are erupted, cool down, and are magnetized. If the field reversed its polarity, the magnetization of the new basalts will be opposite to that of the previous ones. The continuity of the expansion and the succession of the reversals thus give rise to a sequence of sea-floor bands whose TRM has alternatively normal and reverse polarity. Normal bands cause an anomalous field which is added to the Earth's present field (positive anomalies) (Fig. 6.5), the reverse bands a field that is subtracted (negative anomalies). The floor of the oceans is thus comparable to a tape that records the history of the Earth's magnetic field.

The model must clearly be valid on a planetary scale, because ridges occur in the various oceans and polarity reversal is a planetary phenomenon. Its constraints, therefore, are very strict:

- The sequences of maxima and minima on the sides of different ridges must correlate. The width of individual bands may differ from one ridge to another, since it also depends on the spreading rate of each ridge, but the positive/negative sequence must be the same (Fig. 6.6).
- The polarity of a basalt band must be consistent with that of sedimentary and magmatic rocks of the same age sampled on land or as piston cores from the sea bottom. The age of the sea-floor basalts can be established either with isotopic dating or by means of the fossils in the sediments immediately overlying or inter-layered with the lava flows.

Cross-checkings between different ridges and with uplifted sections confirm the validity of the Vine-Matthews-Morley model. So far, we have not discussed the magnetic characteristics of the sea-floor rocks. The model proposes the TRM of basalts as the cause of the anomalies and requires that $J_{TRM} \gg J_i$, i.e. that the contribution of the magnetization induced by the present field would be negligible. Experimental data confirm the hypothesis, since analyses on samples obtained from cores show that the Königsberger ratio of oceanic basalts is always high, even higher than $Q \approx 100$. This figure must be integrated with the structure and composition of the oceanic crust. In the first models, the magnetization source was placed in Layer 2, constituted by basalts with a thickness in the order of 2 km (Fig. 6.2). Today, the structure of the crust is described in greater detail (Fig. 6.7): the pillow lavas of the upper part of Layer 2 (often called in the literature Layer 2A) gradually change with depth to dolerite sills and sheeted dikes (Layer 2B); the gabbros of Layer 3 are thought to be isotropic and possibly in green-schist facies in the upper part (Layer 3A), cumulitic in the lower (Layer 3B). The main source of the anomalies are the pillow lavas, with a thickness of 0.5–0.8 km and a magnetization in the order of 5 A m^{-1}. However, a non-negligible contribution is also brought by dolerites, dikes and isotropic gabbros, with magnetizations in the order of 0.5–1 A m^{-1}. An additional possible contribution could come from serpentinized lherzolites at the Moho depth. Laboratory analyses on remanent magnetization have shown a far more complex picture than the hypothesis of a TRM acquired by rapid cooling of pillow lavas. The Ti-magnetite of oceanic basalts has a high Ti content and thus a low Curie point (<250 °C). As mentioned in Sect. 2.4.1, it tends to oxidize into maghemite, characterized by chemical magnetization (CRM), with

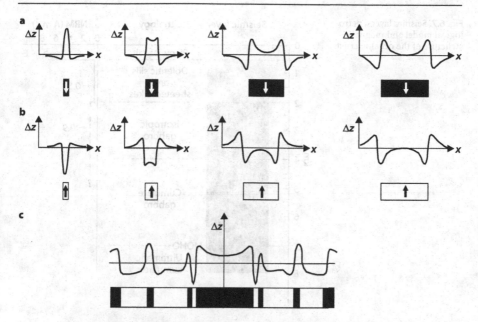

Fig. 6.5. Anomalies of magnetized crustal blocks in the northern hemisphere (vertical remanence for sake of simplicity); **a,b** individual blocks. Normal polarity magnetization (*black block*) causes an anomaly positive above the blocks, negative at the sides; reverse polarity magnetization (*white block*) causes an anomaly negative above the center and positive at the sides. The shape of the anomaly depends on the block's width; **c** overlap of the anomalies of contiguous blocks

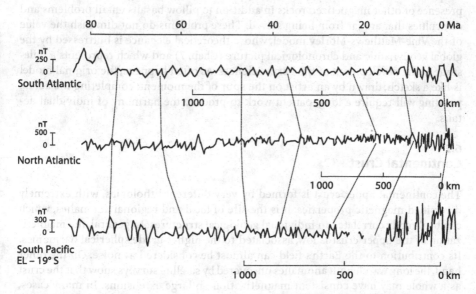

Fig. 6.6. Correlation of magnetic anomalies profiles across the South Atlantic, North and South Pacific ridges. The position of the ridge axis is given by the origin of the time scale (0 Ma) in *the upper part* of the figure. The different length scale of the profiles takes into account the different spreading rate (from Menard 1986, simplified after Heirtzler et al. 1968)

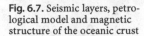

Fig. 6.7. Seismic layers, petrological model and magnetic structure of the oceanic crust

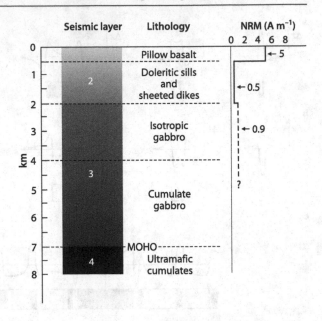

higher Curie point and greater stability. However, this process entails that the CRM is no longer strictly synchronous to the emplacement of the lavas. Therefore, the resulting picture is highly complex: clearly, the hypothesis of TRM magnetized blocks is simplistic and both the physicochemical processes influencing remanence and the presence of other magnetized rocks in addition to pillow basalts entail problems and difficulties that are far from being solved. These problems do not diminish the value of the Vine-Matthews-Morley model, whose theoretical elegance is buttressed by the global geodynamic and chronological picture (Chap. 7) and which represents a milestone in the development of the Earth Sciences. As often happens, the original model is like a sketch, drawn by an artist on the spur of the moment: completing the actual painting will require a long, patient work to produce the harmony of individual details.

6.2
Continental Crust

The continental upper crust is formed by very different lithologies, with extremely diversified magnetic properties. It is the site of local and regional anomalies, which can be directly correlated to individual geological structures. In general, the magnetization of the upper crust is low, associated to the higher-grade spherical harmonics. Its contribution to the Earth's field can almost be considered as noise. On the other hand, the long wavelength anomalies uncovered by satellite surveys show that the crust as a whole may have consistent magnetization on large extensions. In many cases, mathematical interpretation requires crustal blocks several kilometers thick to have a magnetization in the order of 2 to 10 A m^{-1}. Given the heterogeneity and low magnetization intensity of the more superficial rocks, strongly magnetized rocks must occur at greater depth, presumably in the lower crust. The lower crust is constituted prevalently by rocks in amphibolite and granulite facies; quite high magnetite contents can be, but not always are, found in granulites. In addition to magnetite content, the anomalies depend on the thickness of the source bodies. The depth of the bottom of the granulite layer depends on the geothermal gradient, because temperature drives the granulite $\rightarrow$ eclogite phase transition: magnetite disappears and Fe is transferred in paramagnetic, silicate minerals such as garnets. In cratonic regions, the thickness of the magnetized layer can be considerable, because the crust is thicker than elsewhere and the heat flow is low, and thus the Curie isotherm is deeper than usual. Systematic measurements on samples from sections of the lower crust brought to the surface by tectonic processes (such as Lofoten-Vesterålen in Norway, Kapuskasing and Pikwitonei in Canada, Ivrea-Verbano in the Alps) have shown, however, that the lower crust petrology is highly articulated, there is no general correspondence between petrographic facies and magnetite content, and, with few exceptions, the values of the magnetization intensity are systematically lower than those required to cause the anomalies. Just to have an idea, an induced magnetization of 2 to 10 A m^{-1} means a susceptibility of $\kappa \approx 10^5$ µSI, which in turn corresponds to a magnetite content of around 10%, quite exceptional for the rocks of the continental crust. In the absence of a petrologic hypothesis that could explain the presence of a high magnetite content, it has been supposed that in the physical conditions of the lower crust the remanent magnetization intensity will grow. However, this hypothesis is not supported by experimental data, which suggest only the possible presence of a thermal VRM: at high temperatures, in the order of 400–500 °C, grain magnetization is softer than at ambient temperature, promoting the acquisition of a VRM parallel to the present field. The consistency of the direction within different geologic bodies would result in a magnetized layer of regional extension

In conclusion, the hypothesis that magnetite is the source of the magnetization in the lower crust has no enough experimental support: the term *missing* magnetization has come to be used to indicate the magnetization of the continental crust that must be there, because it causes the anomalies, but whose cause is unknown. New experimental tests, instead, indicate that the solution may not be found because it is being sought in the wrong place. The most common ferromagnetic minerals in rocks are magnetite and hematite, and magnetite has always been considered the main source of the anomalies, because its magnetization is greater than that of hematite by two orders of magnitude. However, recent studies have shown that this assumption is not always true. In the case of MD grains, the movements of the walls in the presence of an external magnetic field are hampered by the self-demagnetizing field (Chap. 2), which in hematite is about 1 000 times smaller than in magnetite. Hematite then reaches the TRM saturation in much lower fields than those necessary for magnetite. The Earth's field, though weak $F \approx 0.05$ mT, is near the saturation field of MD hematite, which can so acquire a TRM ≈ 20 times greater than that of MD magnetite. Another hypothesis stems from the magnetic characteristics of metamorphic and igneous deep crustal rocks that have a high hemo-ilmenite content and have cooled very slowly. Their NRM can have intensity in the order of 10 A m^{-1}, is highly stable and extremely resistant to Af and thermal demagnetization. These characteristics cannot be associated to MD magnetite, which is nonetheless present in these rocks. Slow cooling promotes the exsolution of hemo-ilmenite crystals, with the pervasive formation of lamellae, whose thickness can be as low as 1–2 nm. The presence of thin lamellae of (paramagnetic) ilmenite in a host crystal of (antiferromagnetic) hematite causes an imbalance between the equal and opposite magnetizations of the two sub-lattices of hematite (Fig. 6.8), resulting in a net ferromagnetic moment associated to the lamella, which retains the coercivity and thermal stability typical of hematite. The net moment of a crystal is all the more intense the thinner, more numerous the lamellae are. This type of remanent magnetization, called lamellar magnetism, has been proposed as the source of the intense anomalies associated to the banded iron formations, metamorphic rocks that derive from sedimentary deposits exclusive of the Proterozoic, in which the abundance of Fe was linked to the particular characteristics of the Earth's primitive atmosphere.

While the origin of the long wavelength anomalies is still unknown, their planetary distribution suggests that many of them pre-date continental drift. It is thus plausible that they mark major tectonic processes that led to the formation of the continents' cratonic cores and their eventual assembling in the Pangea supercontinent.

Fig. 6.8. Model of hemo-ilmenite multilayer lamellar magnetism in a hematite host crystal. Symbols: *1* = ilmenite Ti layer; *2* = ilmenite Fe layer; *3* = hematite Fe layer; *4* = contact layer; The *black arrows* indicate the direction and magnitude of the magnetic moment of each cation layer within the antiferromagnetic hematite host. The position of the paramagnetic ilmenite lamella determines the direction of the associate moment (*open arrow*); **a** the moments of the two lamellae are in phase and sum up to give a net moment; **b** the moments are out of phase and cancel each other (from Robinson et al. 2002)

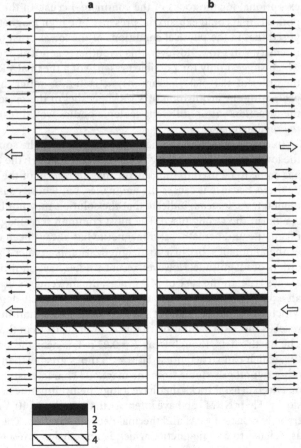

6.3
Global Maps of the Earth's Field

Satellite measurements provide global maps of the Earth's field (Sect. 3.4) which are then processed further with the main challenge of separation of the core and lithosphere contributions. The map of the magnetic field at the CMB is used to study the models of the geodynamo (Chap. 1), while that of the lithospheric field carries valuable geological information at the continental and global scale. Plate 1 shows the map of lithospheric magnetic anomalies at the altitude of 400 km, derived from the POGO and MAGSAT satellites. The maximum peak-to-trough value is 25–30 nT, less than 1‰ of the total intensity actually measured. This figure shows the delicacy of processing satellite data. The anomalies do not show a systematic correlation with ocean-continent boundary, as could be expected based on the different magnetic characteristics of the two types of crust. The missing correlation is traditionally explained by the fact that the wavelength of the expected anomalies is similar to that of low grade spherical harmonics, so they are filtered by the residuation process together with the contribution of the core's field.

The altitude of acquisition of satellite data (Fig. 3.8) is about ten times greater than the thickness of the crust; therefore, the interpretation of the anomalies can not discriminate the magnetic layering of the crust, i.e. magnetization variations with depth. Just to give an idea, if the depth of the source body varies by 10 km, the measured magnetic field varies by 0.5 nT. The basic model for interpretation (SEMM = standard Earth magnetization model) thus considers the crust to be formed by prismatic blocks: magnetization is uniform in each individual block, but laterally variable from one block to another. Each block is characterized by the product of its susceptibility κ times thickness h, indicated with the symbol $\zeta = \kappa \cdot h$ (SI · km). The remanent magnetization of the oceanic crust is not considered, because the wavelengths of its magnetic anomalies are small relative to the satellite's altitude. The sole exception is the crust associated to the Cretaceous Quiet Zone, formed during a normal polarity period which lasted over 30 Myr (Chap. 7), whose width is large enough to cause anomalies detected by the satellites.

The magnetic layering of the crust cannot be derived from the anomalies, but in the case of structures in which crust layering is known from other geophysical and geological methods, the equivalent magnetization of the block can be obtained in greater detail. The crust structure is simplified in layers of mean thickness (Fig. 6.9) and each layer is assigned a value of susceptibility based on the velocity of seismic waves and other lithological evidences. Once the values are assigned to the layers of the block and to those of the surrounding crust, serving as background, their differences yield the susceptibility contrasts $\Delta\kappa$. The whole block's equivalent susceptibility contrast is obtained by computing a weighted average, where the weights are the thicknesses h of the individual layers: $\Delta\kappa = \Sigma\Delta\kappa_i \cdot h_i / \Sigma h_i$. In these forward models, it is clearly possible to consider also layers with magnetization different from that induced by the present field, as may be required to interpret anomalies correlated to rocks with strong remanent magnetization.

The Bangui anomaly, in central Africa, is one of the most prominent anomalies in Plate 1, and historically the first satellite anomaly to be extensively studied. It consists

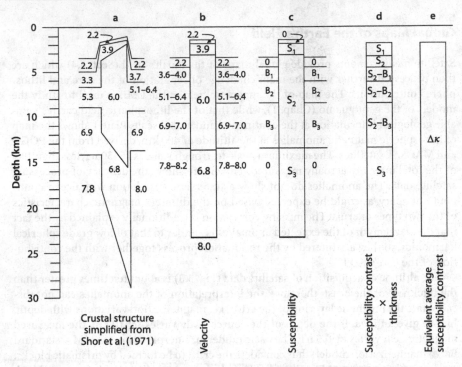

Fig. 6.9. Procedure for calculating the equivalent average susceptibility contrast of a layered crust. Left to rigth: **a** seismic structure, with *P*-waves velocity (km s⁻¹); **b** crustal structure simplified into average thickness blocks and layers; **c** susceptibility assigned to each block of the source body (*S*) and each layer of the background crust (*B*); **d** susceptibility contrast of individual anomalous blocks (*S–B*); **e** the equivalent average susceptibility contrast Δκ is calculated as weigthed average (see text for further explanation) (from Frey 1985)

of a marked minimum (–1 500 nT in ground survey, –1 000 nT in airborne survey flown at 3 km altitude, –20 nT in MAGSAT survey at 375 km altitude) located on the central African craton, flanked by two smaller maxima: to the south, on the Congo Basin, to the north on the central Africa Pan-African belt. Modeling the anomaly requires a source body located at a depth of 3–7 km, with horizontal dimensions in the order of 800 km and with a magnetization of 10 A m⁻¹, mostly of the remanent type. Three hypotheses have been put forward on its nature: a great intrusion of ultrabasic magma with consequent formation of highly magnetic minerals, meteoritic iron associated with a very large impact in early Pre-Cambrian and widespread occurrence of banded iron formations.

Another considerable anomaly in Plate 1 is the one located along the Tornquist-Teisseyre zone, which highlights the limit between the east-European Pre-Cambrian platform and the central-European Palaeozoic platform, whose crust characteristics are very different: the first one is thicker with low heat-flow, the second one is thinner with higher heat-flow. The Tornquist-Teisseyre zone approximately coincides with the zero anomaly line: to the NE, values are positive, to SW, they are negative. The anomaly has been interpreted with two distinct models (Fig. 6.10). The first one hypothesizes two source bodies, one located NE carrying only induced magnetization and one SW,

Fig. 6.10. Tornquist-Teisseyre
zone magnetic anomaly computed at 350 km altitude; contour interval = 2 nT; *full/dashed
contour lines* = positive/nega-
tive anomaly; **a** all source bodies but the small one to
the northeast have remanent
magnetization parallel to the
Permian paleomagnetic
direction for central Europe
($D = 230°$, $I = -15°$).
Figures = DIM in kA. [*DIM*,
depth-integrated magnetization = intensity of magnetization (A m^{-1}) × thickness (m)]
(redrawn after Pucher and
Wonik 1997); **b** source bodies
(both with *DIM* = 30 kA):
A = induced magnetization
($D = 0°$, $I = 65°$), *B* = remanent
magnetization ($D = 180°$,
$I = -50°$) (redrawn after Taylor
and Ravat 1995)

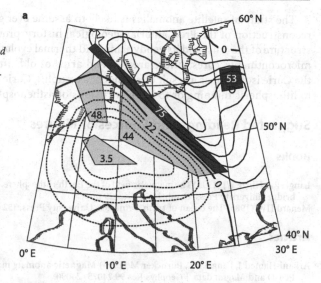

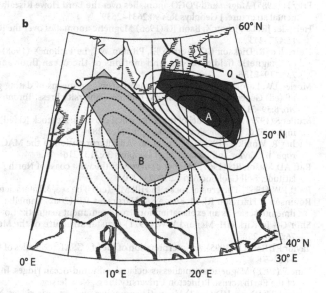

formed of Permian rocks with a strong, reverse-polarity remanent magnetization. The
second one hypothesizes only blocks of Permian rocks with reverse magnetization,
situated along and to the SW of the Tornquist-Teisseyre zone. During the Permian,
the polarity of the Earth's magnetic field remained reverse for about 70 Myr (Kiaman
superchron, see Chap. 7). Given the great extension of Permian rocks in central Europe, the hypothesis, assumed in both models, of consistent reverse remanent magnetization on a sub-continental scale is more than plausible. The differences on the position and the geometry of the blocks reflect the degree of uncertainty that still
characterizes the interpretation of satellite anomalies.

The study of satellite anomalies is likely to assume ever greater importance in the reconstruction of the Pre-Cambrian geological history, providing indications on the structure of the crust and its geodynamic and thermal evolution: correlations between microcontinents' cores, suture areas, failed arms of old rifts; relative movements of the Curie isotherm as a result of subsidence or uplift, Curie isotherm migrations due to lithospheric thinning or to large intrusions of asthenospheric material.

Suggested Readings and Sources of Figures

Books

Langel RA, Hinze WJ (1998) The magnetic field of the Earth's lithosphere. The satellite perspective. Cambridge University Press, Cambridge, UK, 429 pp
Menard HW (1986) The ocean of truth. Princeton University Press, 352 pp

Articles

Arkani-Hamed J, Langel RA, Purucker M (1994) Magnetic anomaly maps of the Earth derived from POGO and *Magsat* data. J Geophys Res 99:24075–24090
Frey H (1985) *Magsat* and POGO anomalies over the Lord Howe Rise: Evidence against a simple continental structure. J Geophys Res 90:2631–2639
Heirtzler JR, Le Pichon X, Baon JG (1966) Magnetic anomalies over the Reykjanes Ridge. Deep-Sea Res 13:427–443
Heirtzler JR, Dickson GO, Herron EJ, Pitman WC, Le Pichon X (1968) Marine magnetic anomalies, geomagnetic field reversals, and motions of the ocean floor and continents. J Geophys Res 73:2119–2136
Morley LW, Larochelle A (1964) Paleomagnetism as a means of dating geological events. In: Osborne FF (ed) Geochronology in Canada. Toronto University Press, Toronto (Roy. Soc. Canada Spec. Publ. No. 8:39–51)
Müller S (1977) A new model of the continental crust. In: Heacock JG (ed) The Earth's crust. AGU, Washington, D.C. (Geophysical Monograph 20, pp 289–317)
Pucher R, Wonik T (1997) Comment on "An interpretation of the MAGSAT anomalies of Central Europe" by Taylor and Ravat. J Appl Geophys 36:213–216
Raff AD, Mason RG (1961) Magnetic survey off the west coast of North America, 40° N latitude to 52° N latitude. Bull Geol Soc Am 72:1267–1270
Raitt RW (1963) The crustal rocks. In Hill MN (ed) The sea, 3. Interscience, New York, pp 85–102
Robinson P, Harrison RJ, McEnroe SA, Hargraves RB (2002) Lamellar magnetism in the haematite-ilmenite series as an explanation for strong remanent magnetization. Nature 418:517–520
Shor GG Jr, Kirk HH, Menard WW (1971) Crustal structure of the Melanesian area. J Geophys Res 76:2562–2586
Taylor PT, Ravat D (1995). An interpretation of the MAGSAT anomalies of Central Europe. J Appl Geophys 34:83–91
Vine F (1968) Magnetic anomalies associated with mid-ocean ridges. In: Phinney RA (ed) The history of the Earth's crust. Princeton University Press, New Jersey
Vine FJ, Matthews HD (1963) Magnetic anomalies over oceanic ridges. Nature 199:947–949

Chapter 7

Magnetic Chronology

The Earth's magnetic field changes over characteristic times which range from a fraction of a second to hundreds of thousands or millions of years (Sect. 1.3). Since the field direction and intensity are recorded by the rocks, their variations over times exceeding some years can be unravelled by measurements of the remanent magnetization (Fig. 4.9). Analysis of rocks of increasing age allows to trace the history of the magnetic field and paleomagnetism can be used as a chronological tool: the polarity, direction and intensity of a rock's remanence are compared with reference curves and the rock's relative age is derived as in the traditional geological chronology (fossil record, isotope curves, etc.). Methods and techniques differ according to the age and thus the time resolution needed; a rough distinction gives

- secular variation (SV, or archaeomagnetism), which covers the historical and prehistorical times, let say the last 10 000 yr;
- paleosecular variation (PSV), which for the time being covers with a satisfactory resolution the upper and middle Pleistocene;
- magnetic stratigraphy, which covers all age ranges of the Earth's geological history.

Magnetic chronology potentially has three enormous advantages over traditional geochronological methods, because the studied signal

1. is synchronous at large scale. Polarity reversals do not depend on geographic location and are synchronous all over the world. Secular and paleosecular variations are consistent at continental scale.
2. has very high resolution power. The time required for a polarity transition is in the order of some thousand years. These are not many if we consider the Plio-Pleistocene and they represent little more than an instant when the rocks' age is hundreds of Ma. The Earth's field direction can vary in the order of $\approx 0.1°$ per year; in the most favorable cases, the resolution power of SV is of a few tens of years.
3. is independent from the facies. All the rocks formed in a given region at a given time recorded the same magnetic field: marine and continental sediments, dikes, lava flows, intrusive stocks, etc.

7.1
Geomagnetic Polarity Time Scale (GPTS)

The basic principle of magnetic stratigraphy is simple. The Earth's magnetic field reversed its polarity from time to time and the sequence of the reversals was recorded in the rocks: rocks of the same age recorded the same polarity. On the one hand, the paleomagnetic study of rocks of known age allows to reconstruct the time sequence of the reversals (i.e. the reference scale), on the other hand rocks of unknown age can be dated by comparing the sequence of the polarities they have recorded with the reference scale. A robust reference scale is thus the essential condition for magnetic stratigraphy.

The first scales were drawn from volcanic rocks whose TRM was acquired at the time of the emplacement, determined using isotopic dating methods. Volcanism, however, is an intermittent process providing discontinuous records of reversal; moreover the older the studied rock the larger the analytical error on its isotopic age: for ages of a few Ma, the error reaches few 100 ka, hence becomes greater than the duration of many polarity intervals. The first scales date back to the pioneer times of paleomagnetism (Fig. 7.1) and extended back to about 5 Ma; even in the most favorable conditions of volcanic activity distributed over a long time span, as in Iceland, they did not extended beyond 15 Ma.

Ocean floors provide the longest and most continuous record of the Earth's field reversals (Sect. 6.1) and the most accurate frame to construct the reference scale. A profile of oceanic anomalies depends on the recorded sequence of reversals and on the expansion rate of the sea floor. These two phenomena are mutually related by the relationship $w = vt$, where w is the width of an anomaly measured along a profile, t the duration of the corresponding polarity interval and v the rate of expansion. The first step in the construction of the scale therefore consists in identifying the anomalies and evaluating their width. The record along a single profile may be more or less complete, as may the expansion of a ridge have been discontinuous. Therefore, it is necessary to consider profiles recorded through the ridges of the various oceans and to constrain their expansion within the global geodynamic framework. The final result of thirty years of this enormous work is the Geomagnetic Polarity Time Scale (GPTS), which covers the last 84 Myr (late Cretaceous to Recent). The framework to build the GPTS is given by the anomalies across the South Atlantic ridge, whose expansion rate is better constrained. The widths of the anomalies are obtained with three different steps:

- Category I widths are the mean of those obtained from 5 to 9 profiles processed according to the method summarized in Fig. 7.2: each profile is first projected perpendicular to the strike of the anomaly, then continued downwards at a depth between 1.5 and 2.5 km (approaching the source of the anomaly enhances resolution) and lastly deskewed, i.e. appropriately processed to reduce the dependence of the anomaly curve on the shape of the source–body and the local field direction. The zero points of the deskewed profile are considered as the limits of the polarity intervals.
- Category II widths are obtained by stacking segments of individual profiles treated similarly to the previous ones. Figure 7.3 shows the case of the anomalies closest to the ridge: in each profile the vertical bars represent the limit of the polarity intervals and the stacking of the curves provides the interpretative profile.

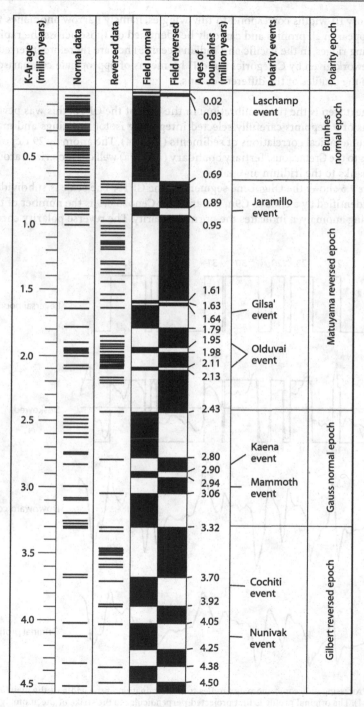

Fig. 7.1. Time scale for polarity reversal in Pliocene-Pleistocene volcanic rocks. Each horizontal bar corresponds to one volcanic unit (from Cox 1969)

- Category III widths correspond to tiny wiggles, namely narrow anomalies that do not appear in all profiles and can only be identified on those across the rapidly expanding ridges, in the Pacific and Indian oceans. They are therefore inserted in the framework given by Categories I and II by means of appropriate comparisons between the profiles of the different oceans.

The next step is the time calibration: in the case of the GPTS, this was performed by means of 9 tiepoints carefully selected integrating isotopic datings and magnetic and paleontological correlations of sediments (Fig. 7.4). The anomaly 29 tiepoint corresponds to the Cretaceous/Tertiary boundary (≈65 Ma) well characterized around the world thanks to the iridium anomaly.

Figure 7.5 shows the Oligocene segment of the GPTS. The scale is subdivided into chrons, identified by a code as C9n: C stands for Cenozoic, 9 is the number of the corresponding anomaly, n indicates the normal polarity. The reverse polarity chron pre-

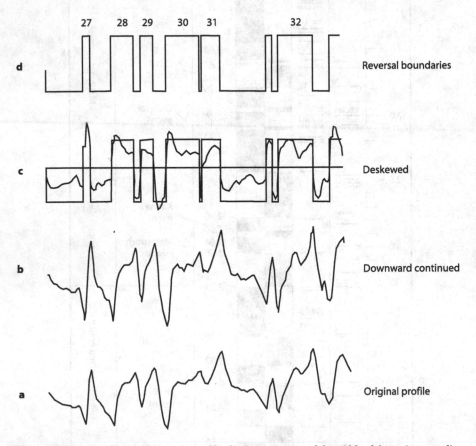

Fig. 7.2. Processing of oceanic anomaly profiles for determination of the width of the main anomalies (category I). The original profile is first projected perpendicular to the strike of the anomaly (**a**), then continued downwards (**b**) and deskewed (**c**). The zero crossings of the deskewed profile give the reversal boundaries (**d**). *Figures* = anomaly numbers (from Cande and Kent 1992)

ceding C9n is labeled C9r. A chron can be subdivided into subchrons. For example, in the chron C10n, four subchrons have been further identified, from the youngest to the oldest: C10n.1n, C10n.1r, C10n.2n, C10n.2r. Cryptochrons have a duration <30 kyr and correspond to the tiny wiggles.

The paleomagnetic data obtained from sedimentary sequences bear a supplementary time constraint to the GPTS: indeed fossils, mainly planktonic foraminifera and nannoplankton, allow to correlate even over great distances the ocean floor magnetic anomalies with the stratigraphic sections derived from ocean piston cores and uplifted

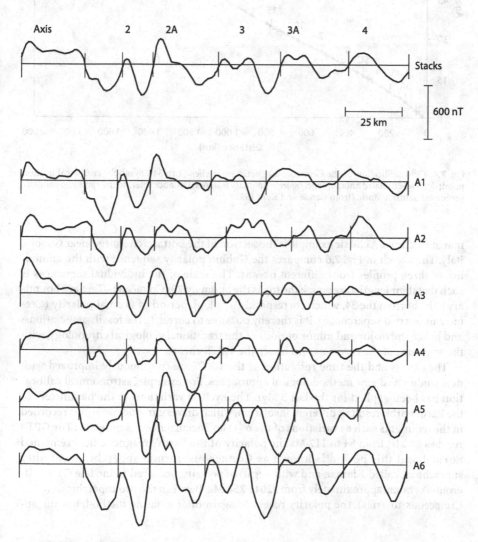

Fig. 7.3. Stacking of oceanic anomaly profiles for determination of the width of the category II anomalies. *Vertical bars across horizontal lines* (= zero level of anomaly profiles) give the reversal boundaries along each profile (*A1* to *A6*) and their average gives those along the stacked profile. *Figures* = anomaly numbers (from Cande and Kent 1992)

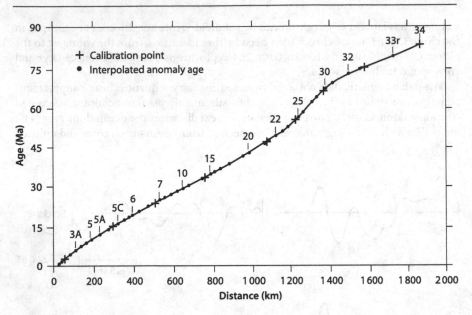

Fig. 7.4. Time calibration of the Cenozoic magnetic anomalies (1 to 34). Symbols: *cross* = calibration point; *dot* = interpolated anomaly age; *figure* = anomaly number. Distance refers to the synthetic anomaly profile for South Atlantic (from Cande and Kent 1992)

marine sections. A classic example is the section of the Bottaccione gorge, near Gubbio, Italy. The sketch in Fig. 7.6 compares the Gubbio polarity sequence with the anomalies of three profiles from different oceans. The scale of the individual sequences is such that their lengths are equal: the top is the anomaly 29 (Cretaceous/Tertiary boundary), the bottom the 34, which corresponds to a long period of normal polarity (Cretaceous normal superchron). It is thereby possible to correlate the fossils associations, and hence the major and minor periods of the traditional geological chronology, with the sequence of the reversals and hence the GPTS chrons.

The details and the time resolution of the GPTS are continuously improved with new studies and new methodological approaches. For example, astronomical calibration has been applied for the last 5 Myr. The cyclical variations in the parameters of the Earth's orbit result in different insolation, which drives climatic variations recorded in the sediments such as variations of the $\delta^{18}O$ and occurrence of sapropels[1]. The GPTS reaches 84 Ma; from 84 to 117 Ma the polarity of the Earth's magnetic field remained normal and this period is known as Cretaceous normal superchron. Another superchron, called Kiaman and with negative polarity, occurred in the late Carboniferous-Permian, approximately from 320 to 250 Ma. Between the two superchrons (early Cretaceous to Trias), the polarity reversals again offer a dating method, but the ap-

---

[1] Sapropels are marine sediments deposited in anoxic conditions and characterized by geochemical peculiarities such as high content of organic carbon, S, and Ba.

Fig. 7.5. The GPTS for Oligocene (23.8 to 33.4 Ma). Absolute age (Ma) is shown in the left column. Symbols: *black* = normal polarity; *white* = reverse polarity; *short bar* = tiny wiggle (modified after Cande and Kent 1992)

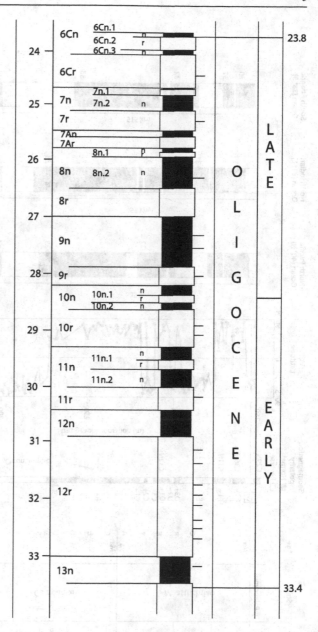

proach is weaker since the magnetostratigraphic scales have mainly been obtained from severely discontinuous uplifted sections. The help of the ocean floor magnetic anomalies is limited to about the interval 0–180 Ma, since the oceanic crust older than middle Jurassic has disappeared in the subduction zones.

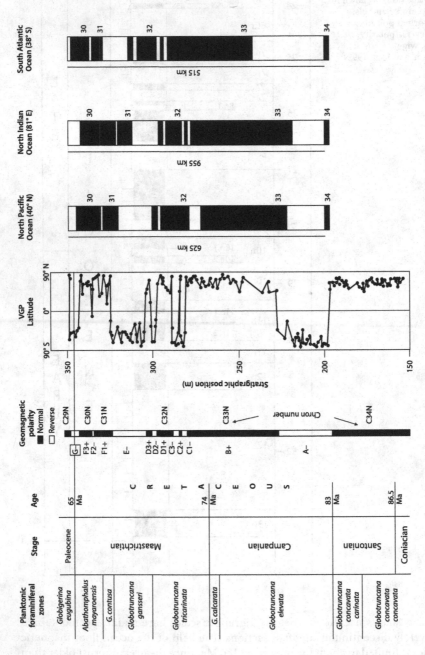

Fig. 7.6. Magnetic stratigraphy in the Bottaccione section (Gubbio, Italy) and comparison with polarity sequence from marine magnetic anomalies; *left:* biostratigraphy; *center:* magnetic zones at Gubbio (*A* to *G*) as derived from VGP latitude, and correlation to the GPTS (*C34N* to *C29N*); *right:* polarity sequences from three oceanic regions (re-scaled to the thickness of the Bottaccione section) (modified after Lowrie and Alvarez 1977)

7.2
Magnetic Stratigraphy

The first point of a magnetostratigraphic study consists in identifying the primary remanence. Laboratory techniques enable to isolate the characteristic remanent magnetization (ChRM, see Sect. 4.3.2) and provide indications on its primary nature. In the case of a sedimentary section, this latter point may not be immediate, because the sampling often concerns various lithologies whose remanence may have different characteristics. A pervasive secondary remagnetization is easy to identify if it involved the whole series, but more difficult if it was restricted to particular lithologies. A successfull reversal test (Fig. 4.23) is a good pro argument to substantiate the primary nature of the ChRM.

Reverse polarities correspond to antipodal ChRM directions: for example, in Pleistocene rocks of the northern hemisphere, the direction $D = 350°$, $I = +50°$ reveals a normal polarity and the direction $D = 170°$, $I = -50°$ reveals a reverse polarity. Since directions are not always so clearly defined, the best practice is to calculate the corresponding VGP and to compare its position with that of the north paleomagnetic pole of the same age. The polarity is considered normal if the two poles are less than 45° far each other, reverse if greater than 135°, transitional in between. In the example above, the North geographic pole corresponds to the Pleistocene paleomagnetic pole: normal polarity corresponds to VGPs falling to the north of latitude 45° N, reverse to VGPs falling south of latitude 45° S, transitional to VGPs falling in between. Once the polarity sequence of the section is obtained, it is compared with the reference scale. It is readily apparent that the comparison makes sense if the section comprises many reversals, as in the above-mentioned case of Gubbio. The reference scale is a Normal-Reverse-N-R-...-N-R series and a succession limited to a few reversals, like N-R-N-R, cannot be univocally correlated to the GPTS. The smaller the stratigraphic interval covered by the section, the more ambiguous the interpretation remains. It is important to keep in mind that the GPTS is a time scale, whereas a section is described by the layers thickness. To pass to the time domain, it is necessary to employ the usual relationship $h = v_s\, t$, where h is the thickness of a layer and v_s the sedimentation rate. A big problem is the fact that v_s is also a function of time, because it varies even within the same section in response to the changing conditions of deposition. The polarity sequence must thus be integrated with independent time constraints, often provided by fossils.

An example of the problems met in the practical application of magnetic stratigraphy is shown in Fig. 7.7, which relates to a section of Pliocene continental sediments in the Guadix-Baza basin (Betic Ranges, southern Spain). The section is about 100 m thick and mainly consists of lacustrine sediments with variable lithology. The samples were collected from 76 distinct levels, spaced by about 1.5 m. About one third of the samples had to be discarded, because the polarity indication was poor. The polarity sequence is formed by eight well defined magnetozones. Two normal polarity levels occur within the thick reverse zone R3. However, they are doubtful: one corresponds to a single site, the other one to low VGP latitudes, to be considered transitional. The correlation with GPTS is enabled by the interpretation of the mammal fauna located in the lower part of the section: magnetozones N1 to N3 can be correlated with the subchrons of the C2An chron (called Gauss epoch in the old terminology in Fig. 7.1,

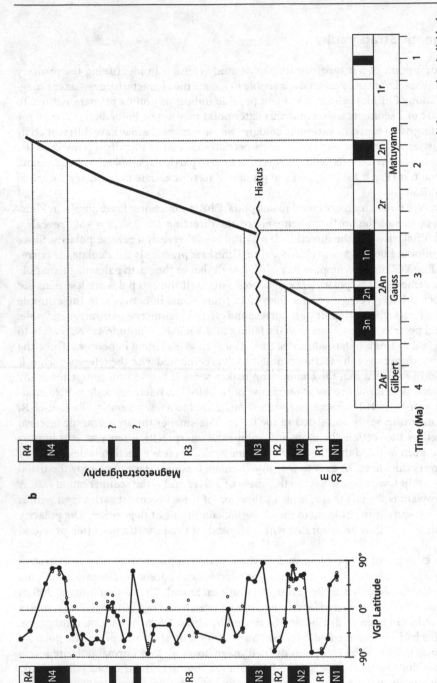

Fig. 7.7. Magnetic stratigraphy in the Galera section (Betic Ranges, Spain); **a** site VGP vs. stratigraphic height and magnetic zones. Symbols: *open dot* = individual sample VGP; *full dot* = mean site VGP; **b** correlation to the GPTS (from Garcés et al. 1997)

still used by Quaternary geologists). The thickness of the N3 magnetozone is smaller than expected from the duration of the C2An.1n subchron. This is due to a hiatus in sedimentation, whose duration is estimated to be ~300 kyr.

Summarizing, the prerequisites for a fruitful application of magnetic stratigraphy are

a a sedimentological study, to trace back the depositional environment and identify any break or discontinuity in sedimentation;
b a paleontological study of the fossils, to provide complementary time constraints;
c a sample spacing close enough to warrant a suitable resolution in time;
d a good quality of ChRM data, to provide reliable polarity in the greatest number of samples.

7.3
Paleosecular Variation

The current polarity chron is C1n, which is also called the Brunhes epoch. The boundary C1r-C1n (Matuyama-Brunhes) is astronomically calibrated at 780 ka. Was this the last reversal? Many younger rocks, both volcanic and sedimentary, yielded reverse polarity. Therefore, it can be admitted that one or more short reverse interval(s) occurred since 780 ka. There is however no general agreement on their number. Their estimated duration varies from 5 to 20 kyr, so their signature is often smoothed out in oceanic sequences with low sedimentation rate, whereas they are often clearly legible in high sedimentation rate marginal sea sequences. Their description in volcanic rocks is wholly random, although few of them have been discovered and studied in details in continental lava flows, and their existence and age have afterwards been confirmed in sedimentary records. Such short intervals of reverse polarity are called cryptochrons, or events in the old terminology. According to some scholars, there is no convincing evidence that they are actual reversals: the reversal process may not have been complete and been reduced to a short excursion of the magnetic poles to the opposite hemisphere, with a rapid return to their normal position. The term *excursion* is thus often preferred to reversal. Their brevity makes their dating difficult: the Laschamp excursion has only recently been dated at 40.4 ±2 ka and the Blake excursion, which is the most reported one, has been dated between 130 and 100 ka, with an estimated duration of about 15 kyr.

One of the various compilations of excursions (some generally accepted, others still controversial) published in the literature (Fig. 7.8) shows that their "frequency" is too low with respect to the time resolution usually needed in the Pleistocene chronology. It is therefore necessary to use variations with higher "frequency", such as paleosecular variation (PSV). However, due to non-dipole field contributions, the PSV curves of D and I may significantly vary from one continent to the other: the low frequency signal due to the main field variations (for example the dipole wobble) may have similar characteristics worldwide, but certainly higher frequency signals are coherent only at a scale of few thousand km. Therefore, correlations on a planetary scale, as with GPTS, are not possible and the reference curves are different from one region of the Earth to another. The most suitable rocks for PSV study are sediments with high sedimentation rate, which can provide a high resolution and a good continuity; in particular:

Fig. 7.8. Summary of reversal excursions during chron C1n (= Brunhes epoch). Excursions are indicated in *bold* (global occurrence and good dating) or in *normal print* (occurrence in limited regions or uncertain) (from Langereis et al. 1997)

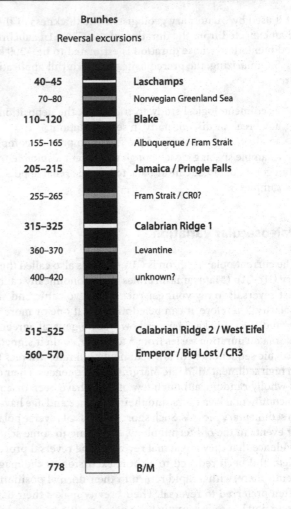

Brunhes
Reversal excursions

40–45	**Laschamps**
70–80	Norwegian Greenland Sea
110–120	**Blake**
155–165	Albuquerque / Fram Strait
205–215	**Jamaica / Pringle Falls**
255–265	Fram Strait / CR0?
315–325	**Calabrian Ridge 1**
360–370	Levantine
400–420	unknown?
515–525	**Calabrian Ridge 2 / West Eifel**
560–570	**Emperor / Big Lost / CR3**
778	B/M

oceanic sediments of continental margin and lacustrine sediments, especially those of the *maar* volcanic lakes, often rich in volcanoclastic magnetite.

When studying the PSV, it is important to check the possible sources of error, whose weight progressively increases as a better time resolution is desired. The ChRM of sediments can be biased by the inclination shallowing and its acquisition is not instantaneous: the lock-in process (Sect. 4.1.3) smoothes the signal and introduces a time lag that is difficult to evaluate. Moreover, lake and marine sediments are sampled as cores often lacking azimuthal orientation: ChRM inclination is absolute (if coring was vertical), but declination is relative. In lacustrine deposits, the time constraints are provided by varve counting, by isotopic dating on organic material or tephra levels and by pollen analysis, which is a great help in correlation both between different lakes and with paleoclimatic data.

The PSV curves of a lake are obtained by retrieving different cores from the bottom. Their lengths corresponds to different time intervals, because the sedimentation rate varies from one point of the basin to the other. If the lake is not very large, the general characteristics of sedimentation remain qualitatively similar; the lengths of each core are brought back to a common depth scale correlating the physical characteristics of the sediment: magnetic susceptibility (Fig. 7.9), dry density, organic matter content, etc. The D, I curves of the individual cores, usually obtained after Af demagnetization, are then analyzed with signal processing techniques: individual curves are merged together by stacking and the merged curve is cleaned of noise and smoothed by filtering. Using available time constraints, the depth scale can be transformed into a time scale along which the D and I variations are plotted providing a merged PSV curve for the lake. Comparison between merged curves of different lakes allows to estimate the reliability of the method and the quality of the results: high degree of reproductibility indicates that the rocks did record the same phenomenon. Age discrepancies are often observed; they result from the accumulation of approximations and errors over the various steps. Differences of a few kyr can be obtained for the interval 20–100 kyr (Fig. 7.10); they demonstrate the accuracy of the method.

Curves from several lakes of the same region, integrated with data from volcanic rocks, concur to construct regional PSV master curves of declination and inclination.

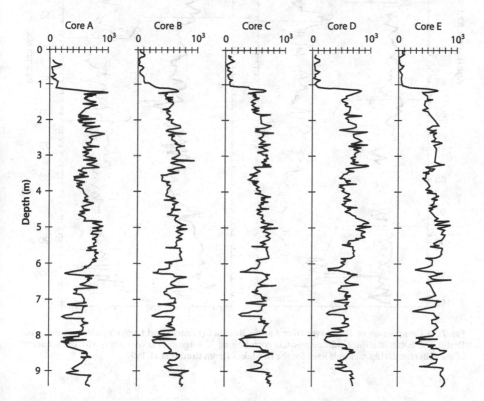

Fig. 7.9. Magnetic susceptibility logs of six bottom cores from Lac du Bouchet (France) transformed to a common depth. Units are 10^{-8} m^3 kg^{-1} (mass suceptibility, χ). *Core D* was used as reference, the other cores were re-scaled correlating first the marker peaks and then extended populations of randomly chosen values (from Thouveny et al. 1990)

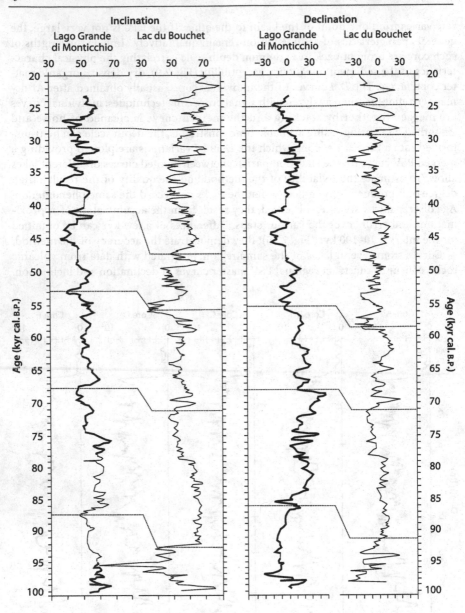

Fig. 7.10. Comparison of *D, I* curves from Lac du Bouchet (France) and Lago Grande di Monticchio (Italy). The *horizontal lines* indicate possible correlations. The age shift is consistent with the accuracy of the different dating methods used for the two lakes (from Brandt et al. 1999)

PSV intensity data on the other hand may be directly compared on a planetary scale (Sect. 4.5.5); they provide relative value of the main field's moment, in the same way as marine sediments.

7.4
Archaeomagnetism

Modern measurements of D, I started in the 17th century and the first absolute measurements of the Earth's field intensity was done in the 1830s. In Europe, the most complete historical series show that over the past four centuries declination has undergone a swing of about 20° and inclination has decreased by 5°–6°. Secular variation (SV) is therefore extremely rapid and can be used to date archaeological findings that acquired a remanent magnetization at the time they were made. Baked clay is the classic archaeomagnetic material: any clay has some iron oxide or hydroxide content, and the hydroxide turn into oxides when heated. Pots and bricks are baked at temperatures as high as 600–800 °C: any remanence they may have is erased and a new TRM is acquired during cooling, aligned with the magnetic field existing at that time.

Pottery and bricks can provide the paleointensity of the field, but they lose the absolute paleodirection because they are removed from the furnace after firing. Only if the firing position is known, inclination can be determined, as demonstrated as early as in the mid 19th century from the observation that the practice of placing the tiles to be fired alternatively upside down, to optimize the use of space inside the furnace, resulted in magnetizations of opposite direction to the shape of the tile. In the case of kilns and furnaces remains, instead, wall bricks are still in place and may provide also the direction of the field. The same holds true for the soil of a site that served as a fireplace. In all these cases, remanence is erased at every firing, to be re-acquired during the subsequent cooling. The measured remanence refers to the last time the kiln or fireplace was used and archaeomagnetism provides an *ante quem* dating, to be interpreted within the archaeological context of a settlement.

Other man-made objects containing ferromagnetic minerals can potentially be used for archaeomagnetism. A wall is built with stone blocks removed from a quarry: their position in the wall is random and their remanence variously oriented relative to the magnetic north. They acquire a VRM whose intensity depends on the time elapsed since the construction of the wall (Eq. 4.10). Powdered hematite has been universally used as a pigment to prepare the red color: after application of the color to a wall and before it dries, the hematite grains can orient themselves according to the magnetic field and the color film, once dried, fixes this direction. A mural thus records the direction at the time it was painted. Even very recent rocks may be used in archaeomagnetism, provided they can be precisely dated. In the sediments of some North American lakes, the pollen of allochtonous plants imported by settlers allows to distinguish the pre- and post-settling sediments. Volcanic rocks of the last millennia can be dated thanks to written sources or with ^{14}C dating of carbonized vegetal remains. A good example of the consistency of the archaeomagnetic method is given by the Roman city of Pompeii, destroyed by the A.D. 79 eruption of Vesuvius. The fine-grained pyroclastic deposits, the lithic clasts they include together with the tiles from the collapsed roofs, the kiln of a lamp maker's shop and a painting in the thermal baths yield archaeomagnetic directions that are indistinguishable from each other (Fig. 7.11).

Archaeomagnetic dating is relative and needs a reference curve constructed by analyzing a great number of findings dated with other methods, such as pottery typology, coins, ^{14}C, written sources, etc. If dating is meant to be very accurate, the reference curve can be used over distances not exceeding a thousand km. Historical mea-

Fig. 7.11. Equal-area projection of archaeomagnetic directions, with α_{95} circle of confidence, from rocks of the A.D. 79 eruption of Vesuvius and archaeological findings at Pompeii

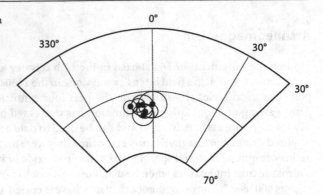

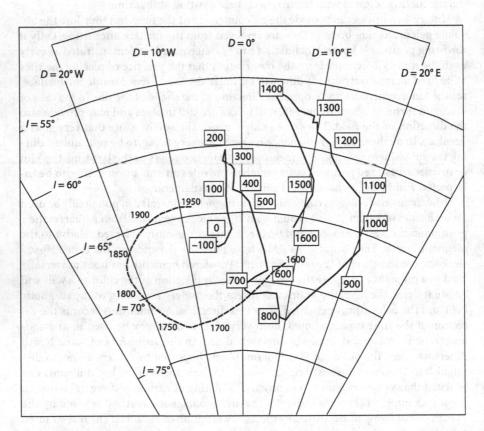

Fig. 7.12. Archaeomagnetic SV curve for France (equal-area projection). Symbols: *dashed line* = historical direct measurements; *full line* = archaeomagnetic data; *box* = A.D. age (from Bucur 1994)

surements and modern ones by geomagnetic observatories demonstrate that SV is different in locations at greater distances. SV curves are presented as stereographic projections (Fig. 7.12), or as *D, I* curves similar to the PSV curves (Fig. 7.13).

Fig. 7.13. Archaeomagnetic SV inclination, declination and intensity curves for Bulgaria (courtesy M. Kovacheva)

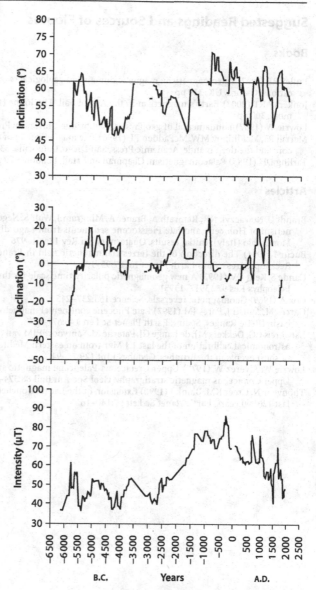

Since the magnetic pole moves around a cap centered on the geographic pole, it is inevitable that the farther one goes back in time, the larger the number of points in which the SV curve intersects becomes. Hence, dating is no longer univocal and it yields two or more distinct ages. This is not a problem for more recent times: in the case of Fig. 7.12, the crossing at the 17th and the 8th centuries can easily be solved within the historical context of the archaeological site. When little or no context is available, as in prehistory, the time resolution of SV inevitably decreases.

Suggested Readings and Sources of Figures

Books

Butler RF (1992) Paleomagnetism: magnetic domains to geological terrains. Blackwell Scientific Publications, Oxford, UK, 319 pp
Jonkers ART (2003) Earth's magnetism in the Age of Sail. The Johns Hopkins University Press, Baltimore, 300 pp
Lowrie W (1997) Fundamental of geophysics. Cambridge University Press, Cambridge, UK, 354 pp
Merrill RT, McElhinny MW, McFadden PL (1996) The magnetic field of the Earth: Paleomagnetism, the core and the deep mantle. Academic Press, San Diego, California, 531 pp
Tarling DH (1983) Palaeomagnetism. Chapman and Hall, London, 379 pp

Articles

Brandt U, Nowaczyk NR, Ramrath A, Brauer A, Mingram J, Wulf S, Negendank JFW (1999) Palaeomagnetism of Holocene and late Pleistocene sediments from Lago di Mezzano and Lago Grande di Monticchio (Italy): Initial results. Quaternary Sci Rev 18:961–976
Bucur I (1994) The direction of the terrestrial magnetic field in France, during the last 21 centuries. Recent progress. Phys Earth Planet Int 87:95–109
Cande S, Kent DV (1992) A new geomagnetic polarity time scale for the late Cretaceous and Cenozoic. J Geophys Res 97:13917–13951
Cox A (1969) Geomagnetic reversals. Science 163:237–245
Garcés M, Agustí J, Parés JM (1997) Late Pliocene continental magnetochronology in the Guadix-Baza basin (Betic Ranges, Spain). Earth Planet Sc Lett 146:677–687
Langereis CG, Dekkers MJ, de Lange GJ, Paterne M, Santvoort PJM van (1997) Magnetostratigraphy and astronomical calibration of the last 1.1 Myr from an eastern Mediterranean piston core and dating of short events in the Brunhes. Geophys J Int 129:75–94
Lowrie W, Alvarez W (1977) Upper Cretaceous-Paleocene magnetic stratigraphy at Gubbio, Italy. III. Upper Cretaceous magnetic stratigraphy. Geol Soc Am Bull 88:374–377
Thouveny N, Creer KM, Blunk I (1990) Extension of the Lac du Bouchet paleomagnetic record over the last 120 000 years. Earth Planet Sc Lett 97:140–161

Chapter 8

Environmental Geomagnetism

The applications of geomagnetism to the environment and human activities are ever more numerous. The first, basic reason is that the mankind's impact on the environment has dramatically increased; the second that magnetism is a fundamental property of matter and the versatility and high analytical ability of magnetic techniques enable every kind of material to be analyzed. Leaving aside the field of biomagnetism, fascinating from the scientific viewpoint and with many consequences in the medical field, we will limit ourselves to the environmental application of the topics we have seen in the previous chapters.

8.1
Environmental Prospecting

Magnetic prospection is a classical surveying technique in archaeology. In its early days, it was mainly used to identify remains to be eventually excavated, as in the 1950s with the tombs of Etruscan necropolis in Tuscany (Italy). Then, it gradually changed to a cost-effective, rapid alternative to excavation. Nowadays, the outcome of a magnetic survey is usually an image of the subsurface which depicts the site and suggests the optimum location of a few test trenches. The very high sensitivity of modern magnetometers makes it possible to detect the small changes in magnetic properties of soils due to biological processes, and thus to construct land use maps which complement the traditional archaeological information. Figure 8.1 shows the magnetic anomaly map in the archaeological site of Monte San Vincenzo, in southern Italy. The result appears the typical one of magnetic archaeological prospecting. A Neolithic village is identified: a large external double ditch encloses small circular ditches and structures probably related to huts. However, numberless point-like anomalies are readily apparent, arranged in a regular grid formed by two orthogonal alignments and superposed to the Neolithic structures. Although today all traces have been lost, the grid represents an olive grove dating back to the ancient Roman pratice of *centuriatio*[1]. Olive trees can live even several centuries, more than enough time for chemical and biological processes in the roots area to leave a mark in the surrounding soil.

The search for intentionally buried artifacts is a typical environmental application of magnetic prospection. The most common case is the unauthorized burial, in sites

---

[1] *Centuriatio* was the distribution of conquered lands to the legionaries, with the dual aim of giving them an award and colonizing newly acquired territories.

Fig. 8.1. Monte San Vincenzo (Puglia, southern Italy). High-resolution magnetic mosaic displayed in 256 gray tones (*black* = –18 nT, *white* = +18 nT). Traces of Roman olive-grove (*regularly spaced dots*) overlay a Neolithic settlement, bounded by a large, double ditch (courtesy M. Ciminale, marci@geomin.uniba.it)

lacking any control, of metal drums containing toxic waste, whose release is destined to pollute surface and ground waters over the years to come. Total field measurements are taken along a closely spaced grid (0.5 to 1 m), often with a magnetometer gradiometric configuration. The higher resolution of gradiometers allows a more precise positioning location which reduces the risk of accidental breaking and waste leakage during removal. Magnetic techniques are continuously improved in test sites, realized by private companies and scientific institutions with the aim of doing direct experiments of the various shallow geophysics techniques used to find buried artifacts and monitor the evolution of contaminant plumes. The map in Fig. 8.2 refers to a test site located in fluvioglacial deposits consisting of conglomerate in a silty-sandy fine matrix (Central Apennines, Italy). Twelve drums were buried in a vertical position with their top at a depth of 4.5 m. The shaded relief total field anomaly map shows a typical dipolar anomaly, with a well defined maximum-minimum axis oriented N-S, parallel to the local direction of the Earth's field. Airborne measurements are often used when the area to be surveyed is large. To avoid the risk of exposure for the operators, airborne surveying is the practice in the case of unexploded ordnance (the so-called UXO).

Another, far more devastating case, is that of land mines. A sure method to identify them has not yet been devised and the ability of magnetic measurements has been strongly reduced, since metal parts have been almost completely eliminated from mines, to make them more difficult to detect. However, a substantial problem in mine clearing stems from *false alarms*, which prolong the operation time and increase the costs. They can be reduced by the coexistence of clues obtained by means of different methodologies and the small magnetic signal caused by the firing pin contributes to a more accurate identification.

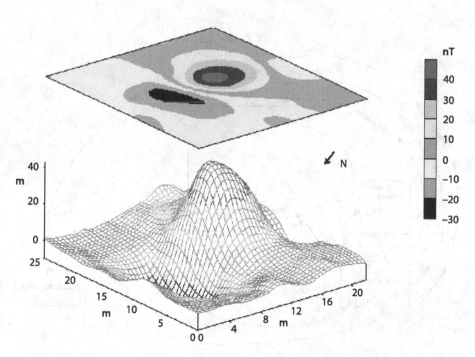

Fig. 8.2. Magnetic anomaly map of a test site for search of buried drums (from Marchetti et al. 1998)

8.2
Enviromagnetic Parameters and Techniques

The need to better understand the magnetic properties of rocks and minerals has over the past thirty years led to a dramatic development in laboratory instrumentation, techniques and methodologies. The field of magnetic measurements has expanded to virtually all materials, including those whose magnetism is so weak that they are hastily dubbed *non magnetic*. In the previous chapters we have seen that very few minerals are ferromagnetic and they always form a small or minimal fraction of a rock. Most of them originate in magmatic and metamorphic rocks, then their grains enter the cycle of disintegration, alteration, transport and final sedimentation in lakes and seas (Fig. 8.3). Ferromagnetic grains constitute a minimal part of the material involved in the processes that lead from the parent rock to a sediment and in the course of this journey they act as tracers. Nowadays, we are able to identify the ferromagnetic grains and reconstruct the way they are connected to the various environmental processes. Environmental magnetism also studies materials of extra-terrestrial origin, such as cosmic dust and micrometeorites, and those produced by human activities and connected with pollution.

An environmental magnetism laboratory is not very different from a paleomagnetic laboratory; the difference is in the way it is used. The materials to be measured are as disparate as they can get: not just rocks, but above all soils, dusts, ice, biological materials, etc. Hence, laboratory procedures must be adapted to particular require-

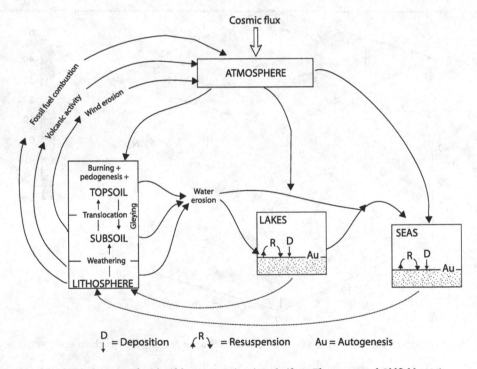

Fig. 8.3. The environmental cycle of ferromagnetic minerals (from Thompson and Oldfield 1986)

ments, case by case. But above all the investigative approach is different. Paleomagnetists are interested in the magnetic field of the past and try to extract its direction and intensity from the rocks' remanence; environmental magnetists are interested in magnetic minerals and try to put them at the right place and time in the environmental cycle, determining their type, quantity and grain size.

Magnetic susceptibility continues to be a basic parameter, but it is expressed in terms of mass, the best to reflect concentration: $\chi = \kappa / \varrho$, where κ is volume susceptibility and ϱ is density. Since κ is dimensionless, the unit for χ is $m^3 kg^{-1}$. Mass susceptibility is often measured at two different frequencies. If the high-frequency value is smaller than the low-frequency one, the material contains grains in the SP state (Sect. 5.2), which may form in the course of various environmental processes. However, when the concentration of ferromagnetics is extremely low, the measured χ may not be significant. For example, the diamagnetic susceptibility of ice masks the signal of the very small quantity of ferromagnetic dust it contains. This quantity cannot be increased by separation, because some hundreds of kilograms of ice would have to be melted to obtain a measurable quantity of dust. Moreover, the measurement of susceptibility does not discriminate the signals, and hence the content, of ferro- and antiferromagnetic minerals, whose environmental significance can be very different. These problems are overcome by measuring the remanent magnetization artificially given to a sample, which can be either isothermal (IRM) or anhysteretic (ARM). A mixture of magnetite and hematite grains produces an IRM acquisition curve dominated by the magnetite signal (low coercivity), but that in fields of the order of 1 T does not yet reach the saturation value (SIRM), due to the high coercivity of hematite (Fig. 8.4). Similar information is provided by the parameter called S-ratio, $S = J_{back} / SIRM$,

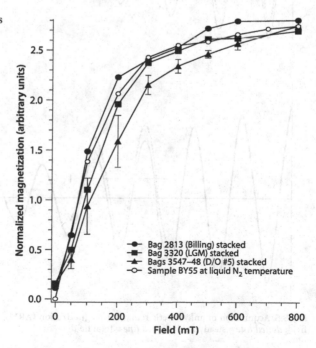

Fig. 8.4. IRM acquisition curves of Greenland ice samples from three different climatic stages and a sample of loess (*BY55*). All samples measured immediately after immersion in liquid nitrogen. Samples from *Bags 3547-48* show the highest content of high-coercivity minerals (from Lanci et al. 2004)

which is obtained by first saturating the sample in one direction (SIRM) and then applying in the opposite direction a backfield with a magnitude in the order of 0.3–0.4 T, enough to saturate the soft magnetite grains but not the hard hematite grains. The remanence J_{back} measured after the backfield will have a value approaching the SIRM (and hence $S \approx 1$) if it is mainly carried by soft minerals, lower (and hence $S < 1$) if the hard ones comprise a significant fraction. The intensities of IRM and SIRM are usually normalized in terms of mass, and hence expressed in A m^2 kg^{-1}, like that of the ARM.

The anhysteretic remanent magnetization (ARM) is the magnetization acquired by a sample subjected simultaneously to an alternating magnetic field, whose initial intensity H_{peak} is progressively reduced until it disappears, and a weak steady field, whose intensity H_{bias} is similar to that of the Earth's field (Fig. 8.5). The alternating field un-

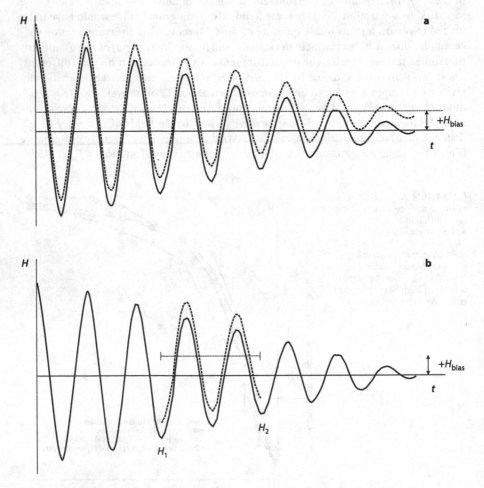

Fig. 8.5. Acquisition of anhysteretic remanent magnetization (ARM). Symbols: *full line* = alternating field; *dotted line* = steady field; *dashed line* = total field

Fig. 8.6. Relationship between ARM susceptibility and magnetic susceptibility for magnetite grains of different size (given in μm) (from Evans and Heller 2003)

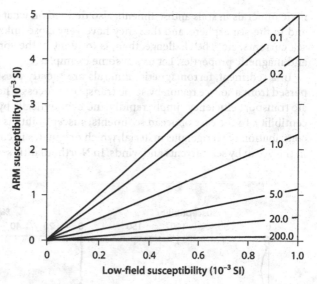

locks the domains with remanence coercivity below H_{peak}, the steady field destroys the symmetry of the alternating field and creates a preferential direction of magnetization. The magnetization J_{ARM} is parallel to the steady field and proportional to its magnitude. Applying the bias field over an interval of values of the alternating field $H_1 > H_2$ (Fig. 8.5), the ARM is acquired only by the grains whose remanence coercivity falls in that interval, $H_1 \geq H_{cr} \geq H_2$. ARM susceptibility, defined as $\chi_{ARM} = J_{ARM} / H_{bias}$ and expressed in $m^3 kg^{-1}$, does not depend on the value of the steady field and it is useful in discriminating the dimensions of the grains. The ARM susceptibility of grains in the order of 0.1 μm is several times larger than that of grains with dimensions of 1 μm (Fig. 8.6).

Another commonly used type of measurement to study grain size is the hysteresis cycle (Sect. 2.2.2), from which we obtain the values J_s, J_{rs} (saturation magnetization and saturation remanence) and H_c, H_{cr} (coercive force and coercivity of remanence). The values of their ratios are indicative of the state of the grains: for SD magnetite $J_{rs}/J_s = 0.5$, $H_{cr}/H_c = 2.0$; for MD $J_{rs}/J_s = 0.02$, $H_{cr}/H_c = 5.0$; for PSD values are intermediate. If the grain population is an SD + MD mixture, measured values fall within the range of the PSD grains. For the hysteresis cycle, as for other types of measurements, new techniques are therefore being perfected which enable one to separate the contributions of the different types of grains and evaluate their percentage content.

8.3
Magnetic Climatology

In Chap. 7, we mentioned the possible relationships between the geomagnetic and the astronomical clocks, derived through comparison of the changes in the magnetic field with those of the Earth's orbital parameters. These parameters also affect solar irradiance and thus the climate. The quantity, type, and degree of alteration of ferromagnetic minerals in soils and sediments also depend on what occurs in the hydrosphere and in the atmosphere, and thus they have very close links with temperature, winds, sea currents, etc. The challenge, then, is to identify the correlations between climate and magnetic properties. Let us see some examples.

In a sediment, ferromagnetic minerals are mostly present as detrital grains, dispersed from a source region by some transport process. Their content allows to evaluate transport efficiency simply, rapidly and economically by measuring magnetic susceptibility. In the case of ocean sediments, susceptibility is strongly influenced by the contribution of terrigenous material, which originates in continents and is distributed in the ocean by sea currents and winds. In North Atlantic sediments, layers with a sud-

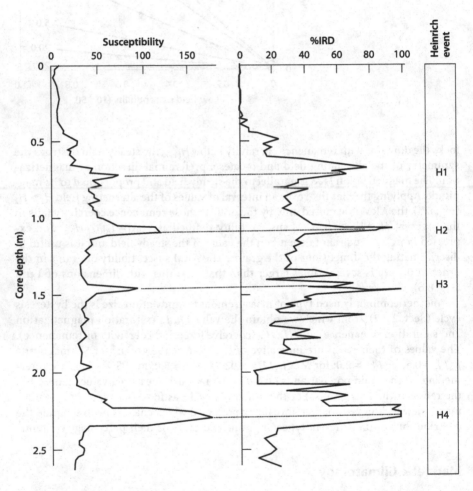

Fig. 8.7. Whole-core magnetic susceptibility (in 10 μSI-units) and ice-rafted detritus (IRD) content (percentage of lithic grains > 150 μm) in North Atlantic sediments. The maxima of the curves correlate with the Heinrich events (from Robinson et al. 1995)

den increase in coarse-grained (>150 μm) terrigenous content, and hence susceptibility, have been identified (Fig. 8.7). This coarse material is called ice-rafted detritus (IRD): it is enclosed in the ice that originates on a continent and reaches the ocean, where it is spread by the currents, melts and finally releases the detrital grains, which can thus be deposited even a great distance away from the coasts. These IRD-rich layers have led to the definition of the so-called Heinrich events, corresponding to the glacial maxima that occurred over the last few tens of thousand years: larger volume of continental ice $\to$ larger quantity of ice transported in the ocean $\to$ higher percentage of IRD $\to$ higher magnetic susceptibility. The magnetic scanning obtained by passing a core inside a coil at the rate of a few cm/minute provides the susceptibility log, from which the layers with greater IRD content are easily identified, leaving the core undisturbed for all further analyses.

A completely different transport mechanism produces something similar in continental aeolian sediments like loess. Normally, wind is not a very efficient means of transport for ferromagnetic grains, with their high densities on the order of $5\,000\,\mathrm{kg\,m^{-3}}$. In periglacial regions, wind circulation is reinforced and a stronger wind can lift and transport farther a larger quantity of ferromagnetic grains. Colder climatic conditions thus correspond to a susceptibility increase. In a section along the course of the river Biya, a tributary of the Ob in southern Siberia, the timing of the susceptibility maxima correlate well with the Heinrich events of the North Atlantic. On the other hand, a simple parameter like magnetic susceptibility is not very selective, so it is essential to determine what its actual meaning is, and this can only be done considering the climatic processes of each region as a whole. The Chinese loess plateau is characterized by alternating layers of loess and paleosols formed by pedogenetic alteration. The layers that maintained the original loess characteristics correspond to cold glacial periods, the pedogenized ones to warmer interglacial periods. The correspondence between susceptibility and climate conditions is opposite to that of the Siberian loess: the χ values are lower in cold periods, higher in the warmer ones (Fig. 8.8). In Chinese loess, hematite is the major ferromagnetic mineral, associated with very minor quantities of magnetite and maghemite. Susceptibility increase is linked to pedogenetic formation of magnetite grains which considerably increase susceptibility, since the χ value of magnetite is two orders of magnitude higher than that of hematite. Other measurements (IRM, Curie temperature) confirm that the new mineral is magnetite and χ variations as a function of frequency indicate the small size of the grains, at the boundary SP to SD.

Lake sediments are an excellent climate archive, often complemented by the chronological information provided by the PSV data, as in the case of the Lac du Bouchet (Sect. 7.3). In this case, magnetic susceptibility correlates well with the alternation of glacial and interglacial periods. In glacial periods, the greater cryoclastic contribution increases the terrigenous content of the sediment and the lower organic productivity limits dissolution and alteration of the ferromagnetic grains. Susceptibility values, therefore, are higher, whereas they are lower in sediments deposited in interglacial periods, characterized by a smaller content of terrigenous material and greater organic productivity. Also in this case the information provided by magnetic susceptibility must be substantiated by other measurements that take into account the com-

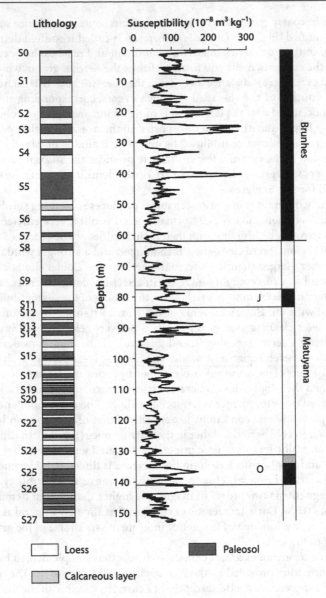

Fig. 8.8. Lithology, mass susceptibility and magnetostratigraphy of the loess/paleosol sequence at Lingtai, central Chinese Loess Plateau. The lithology is well reflected by magnetic susceptibility, which shows enhanced values in paleosol horizons due to neo-formation of magnetic minerals during soil development. The susceptibility profile is based on some 6000 samples, ensuring an almost continuous stratigraphic coverage (courtesy S. Spassov)

plexity of the situation: for example, a strongly reducing environment favors the dissolution of the original detrital magnetite and the formation of greigite.

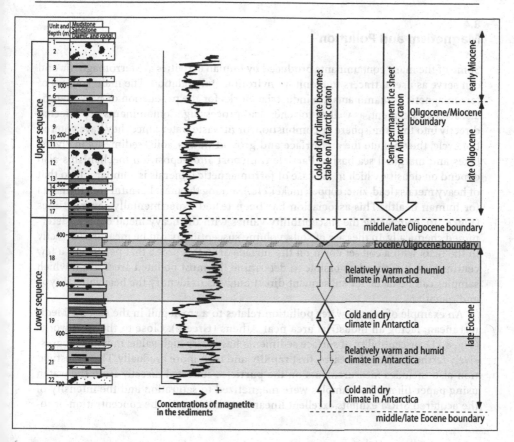

Fig. 8.9. Correlation between magnetite content and climate conditions in Eocene to Miocene sediments from the CIROS-1 core from Ross Sea, eastern Antarctica (courtesy L. Sagnotti)

The previous examples refer to the last glacial periods, but analysis of marine sediments allows to extend the paleoclimatic record farther back in time, albeit with less resolution. Investigation of sediments along the coasts of Antarctica (Fig. 8.9) provides valuable indications on the formation and oscillation of its ice cap, whose extension is one of the main factors controlling the planetary climate of the Earth. Cold and dry climate was not established in Antarctica until the Eocene/Oligocene boundary, with major ice-sheet growth occurring at the early/late Oligocene boundary. Earlier cold intervals indicate that climate had begun to deteriorate in the middle Eocene and late Eocene was a transitional period characterized by repeated alternation of relatively warm/humid and cold/dry conditions.

Magnetic climatology and chronology have become an essential part of the interdisciplinary approach to the study of the late Pliocene to Pleistocene paleoclimate. The climate signature is present in virtually all sediments of this age, since it has not yet been smoothed or erased by later processes such as diagenesis and metamorphism.

8.4
Magnetism and Pollution

Some of the many contaminants produced by human activities are ferromagnetic and
can serve as useful tracers to monitor environmental pollution. The main sources of
ferromagnetic contaminants are industrial works for the production of base materials, steel and cement, as well as for chemical processing. Contaminants are injected
directly into the atmosphere by combustion, or in waste water. Once they have entered
the cycle, they pollute the air, surface and ground waters, soils, sediments in rivers,
lakes and marginal sea basins. Particle transport and deposition mechanisms also
depend on density, which in the case of ferromagnetic minerals is comparable to that
of heavy metals (lead, zinc, copper, nickel, etc.), among the most harmful contaminants
for human health. This association has been tested experimentally several times
(Fig. 8.10), comparing the susceptibility values with the heavy metal content determined chemically. In the case of soils, volume susceptibility can be measured directly
in the field, with a coil set down on the surface or with a probe that penetrates a few
centimeters. It is thereby possible to determine the most polluted areas, from which
samples can be taken for subsequent direct analysis to identify the heavy metal type
and quantity.

An example of coastal water pollution relates to a small gulf in the eastern Mediterranean, facing an industrial area near Athens (Greece). Close to the steel works
(Fig. 8.11) susceptibility of surface sediments has a very high value; moving progressively farther away, it decreases, first rapidly and then more gradually. This investigation also includes measurements on the particles collected directly from the water,
using paper filters. The samples were magnetized to saturation and the intensity of
the acquired SIRM exhibits excellent linear correlation with the concentration of to-

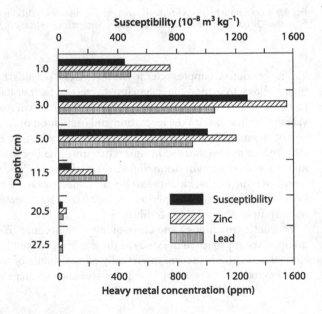

Fig. 8.10. Mass susceptibility, lead and zinc content in a soil near Jaworzno power station (Poland) (from Heller et al. 1998)

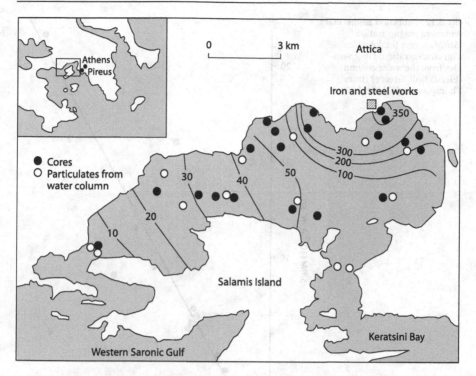

Fig. 8.11. Marine particulate pollution in the Elefsis Gulf (Greece). *Contour lines* = mass suceptibility of surface sediment in 10^{-8} m^{3}kg^{-1} (from Thompson and Oldfield 1986)

tal iron particulates (Fig. 8.12). In this case, too, a simple magnetic measurement provides a first indication of the level of pollution.

The study of atmospheric pollution is particularly important both because the atmosphere is the medium through which many contaminants reach soils and waters, but also because contaminant particles are breathed, reach the lungs directly and can constitute severe health hazards. The highest risk occurs near coal-burning power-plants and in large urban concentrations, above all because of vehicular traffic. The particulate to be analyzed is collected with filters or rubbing surfaces that are directly exposed to air. The first method requires costly installations, the second one does not assure good efficiency. Therefore, the practice of collecting the leaves of trees exposed to pollution is becoming increasingly widespread, as it allows cost-effective, systematic sampling on large areas. Leaves, moreover, provide a measure of long-term pollution, more useful to evaluate human health hazard than the few-days reading from air filters. The leaves of *Quercus ilex*, an evergreen tree very common in Mediterranean regions, have a lifespan up to three years and the content of pollutants they accumulate has been shown to be a function of time. Even more common in urban areas is *Platanus* sp., a deciduous species whose leaves have a lifespan of a few months. A case-history from Rome is shown in Fig. 8.13. The leaves' magnetic susceptibility varies as a function of the distance of the tree from the main roads and railways: in suburban

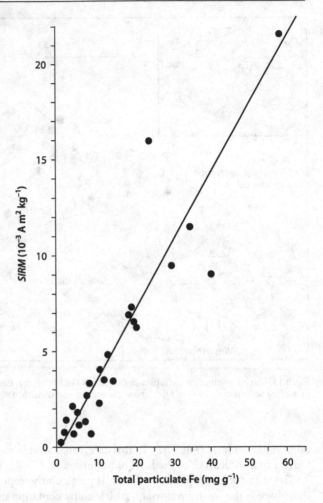

Fig. 8.12. Saturation isothermal remanent magnetization (*SIRM*) versus total particulate iron concentration. Filter samples from the water column (Elefsis Gulf, Greece) (from Thompson and Oldfield 1986)

parks, its values are about 2×10^{-8} m^3 kg^{-1} for both species, where vehicle traffic is most intense they increase up to 10×10^{-8} m^3 kg^{-1} for *Platanus* sp. and 50×10^{-8} m^3 kg^{-1} for *Quercus ilex*.

8.5
Seismo- and Volcanomagnetism

In seismic and volcanic regions the Earth's internal energy is released as elastic and thermal energy conveyed by the rocks. Both kinds of energy interfere with the rocks' remanence, since they cause piezo (PRM) and thermal (TRM) remanent magnetizations. Changes in the regional stress or thermal field will therefore change the remanence of large rocks' volume, which in turn will produce a small magnetic change of the Earth's field. Continuous magnetic monitoring can thus contribute to the surveillance and possibly help in prediction.

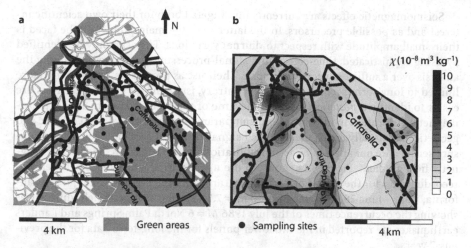

Fig. 8.13. Magnetic susceptibility of leaves of *Platanus* sp. in an urban area in Rome; **a** city map. Symbols: *dot* = sampling site; *full line* = main road; *dashed line* = railway; **b** mass susceptibility map. Contour interval 2×10^{-8} m^3 kg^{-1} (from Moreno et al. 2003)

A number of early works of the 19th century interpreted magnetic time variations in seismic regions as a consequence of earthquakes. Milne, one of the fathers of modern seismology, showed that in many cases this attractive connection was spurious and only due to inertial effects on suspended magnets. Only after the 1960s, with the introduction of absolute magnetometers, proton precession and optical pumping in particular, and noise reduction techniques, the magnetic data recorded during a seismic event could be considered trustworthy and small changes recorded in the magnetograms related to seismic activity. These changes were associated to stress variations in the rocks and Nagata, one of the founders of rock magnetism, introduced the general term 'tectonomagnetism'. The term 'seismomagnetism' is used in case the magnetic effects are directly associated to seismic events. The actual situation is more complex, because both slow- and long-term variations occur. Slow variations of the magnetic field, on the time scale of weeks or months, are referred to local changes of the stress field, which cause piezomagnetic effects (Sect. 4.1.6). Even if very small, they sum up over a large rock volume and their total effect can be detected. Electrokinetic phenomena, due to build-up of electric currents in the presence of an electric double layer in the crust, are proposed as the cause of more rapid variations, on time scales of seconds to days. The amplitude of the field change ΔF due to the seismomagnetic effects depends on many factors and in general ranges in the order of 0.1 nT or even less. Measurements require a network of magnetometers synchronously operating at different sites, and connected to a central station or magnetic observatory, for example via mobile phone. Here, the differences between pairs of stations are computed, often after having averaged data on an hourly or a daily basis in order to smooth very short time variations. In fact besides tectonomagnetic or seismomagnetic effects, small variations can arise also from other sources related to the planetary magnetic field (Sect. 1.3), mainly: the external part of the Earth's field, non-uniform secular variation and the electromagnetic field induced in the crust and upper mantle.

Seismomagnetic effects are currently investigated both for their own scientific interest and as possible precursors. In the latter case, the main problem to be faced is their small amplitude with respect to diurnal variations. They can only be identified by using sophisticated, time-consuming signal-processing techniques and doing the analysis over a sufficiently long time span. Their use as precursors, if any, is therefore limited to long-term prediction. On the contrary, istantaneous coseismic effects are easier to identify *a posteriori*, because the time of the event is known. Simple piezomagnetic dislocation models based on fault parameters determined from seismic and geodetic data usually match the observed signal, provided care has been taken to ensure sensors are not affected by seismic vibrations and located in regions of low magnetic field gradient. Figure 8.14 shows, in the upper part, the difference between two recording sites for the period of 1 day before and after the Landers earthquake (California, USA – June 28, 1992, magnitude $M = 7.3$); similar magnetic field differences showing the occurrence times of the July 1986 $M = 6$ North Palm Springs and Landers earthquakes are reported in the two lower panels for the long-term data for the previous 7 years.

Monitoring the volcanic activity and prediction of eruptions is another geomagnetic application of environmental interest. Volcanomagnetic effects have amplitudes up to 20 nT, higher than that of seismomagnetic ones, but they are felt over shorter distances, not more than 15–20 km from the volcanic edifice. Their possible sources are numerous and can broadly be grouped according to their characteristic duration. Long-term effects are mainly interpreted as due to thermomagnetic processes. The rocks surrounding a magma chamber are heated and lose a fraction of their remanence; on the contrary, cooling of dikes or intrusions at depth produces new magnetized rocks. Tentative correlations have also been suggested between volcanomagnetism and slow changes in the groundwater circulation within a volcanic edifice, which can produce electrokinetic currents. Piezomagnetism is the principal mechanism of short-term volcanomagnetic effects, as a result of stress redistribution due to dike intrusion and opening and propagation of eruptive fissures. The time resolution of volcanomagnetic signals is very high, as magnetic monitoring is continuous and acquisition of PRM is practically istantaneous. It is thus possible to follow the evolution of the volcanic activity in great detail. In the course of the October 2002 eruption at Mount Etna, two stages of geomagnetic intensity changes have been observed, one associated with the October 26 seismic swarm (Plate 4), the other to the opening of eruptive fissures on October 27. In this last case, the rate of growth of the magnetic anomalies allowed one to estimate that the magmatic intrusion traveled northwards at approximately 14 m min^{-1}.

Nowadays tectonomagnetism is considered as a part of a wider scientific field, which includes tectonoelectric observations as well as the extension of magnetic observations to the various parts of the electromagnetic spectrum, from sub-microhertz to radio frequencies. This discipline is called EMSEV, from Electromagnetic Studies of Earthquakes and Volcanoes. Some high-frequency effects have been observed as anomalous electromagnetic emissions associated to, and also before of, moderate to strong earthquakes. A famous case history regards the Chilean earthquake of May 16, 1960 (magnitude $M = 9.5$). In this case a radio emission at 18 MHz was recorded at

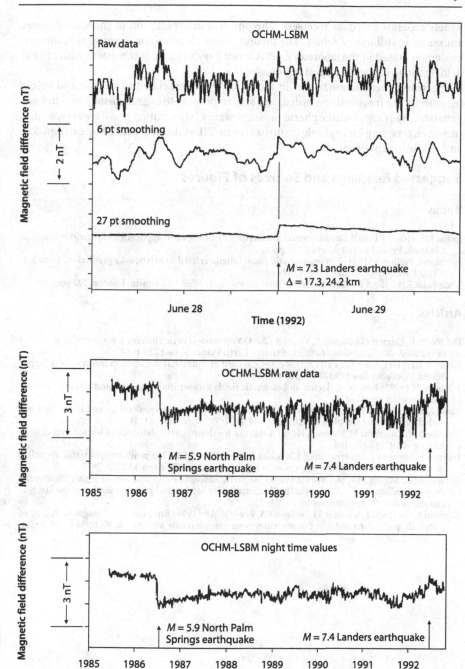

Fig. 8.14. Seismomagnetic signals shown by the magnetic field differences between the recording stations OCHM and LSBM (California, USA) for the Landers and Palm Springs earthquakes (from Johnston 1997)

widely separated, distant receivers. The physical interpretation of these anomalous emissions is still under debate. The theoretical mechanism for waves generation can be demonstrated in the laboratory, yet it is not easy to make it consistent with crustal deformations in the epicentral region.

Electromagnetic disturbances in the ionosphere associated to seismic and volcanic activity are frequently recorded, and able to propagate to great distances. They are generated by trapped atmospheric pressure waves (also named gravity waves, acoustic waves, traveling ionospheric disturbances or TIDs) directly excited by earthquakes and volcanic explosions.

Suggested Readings and Sources of Figures

Books

Evans ME, Heller F (2003) Environmental magnetism. Principles and applications of enviromagnetics. Academic Press, Elsevier Science, 299 pp
Hayakawa, Fujinawa (1994) Electromagnetic phaenomena related to earthquake prediction. Terra Scientific Publishing Company, Tokyo, 677 pp
Thompson R, Oldfield F (1986) Environmental magnetism. Allen and Unwin, London, 227 pp

Articles

Del Negro C, Currenti G, Napoli R, Vicari A (2004) Volcanomagnetic changes accompanying the onset of the 2002–2003 eruption of Mt. Etna (Italy). Earth Planet Sc Lett 229:1–14
Heller F, Strzyszcs Z, Magiera T (1998) Magnetic record of industrial pollution in forest soils of Upper Silesia. J Geophys Res 103:17767–17774
Johnston M (1997) Review of electric and magnetic fields accompanying seismic and volcanic activity. Surv Geophys 18:441–475
Lanci L, Kent DV, Biscaye PE, Steffensen JP (2004) Magnetization of Greenland ice and its relationship with dust content. J Geophys Res 109 (D09104, DOI: 1029/2003JD004433)
Marchetti M, Chiappini M, Meloni A (1998) A test site for the magnetic detection of buried steel drums. Ann Geofis 41(3):491–498
Moreno E, Sagnotti L, Dinerès-Turell J, Winkler A, Cascella A (2003) Biomonitoring of traffic air pollution in Rome using magnetic properties of tree leaves. Atmos Environ 37:2967–2977
Robinson SG, Maslin MA, McCave IM (1995) Magnetic susceptibility variations in Upper Pleistocene deep-sea sediments of NE Atlantic. Implications for ice rafting and paleocirculation at the last glacial maximum. Paleoceanography 10:221–250
Sagnotti L, Florindo F, Verosub KL, Wilson GS, Roberts AP (1998) Environmental magnetic record of Antarctic paleoclimate from Eocene/Oligocene glaciomarine sediments, Victoria Land Basin. Geophys J Int 134:653–662

Chapter 9

Historical Notes

Magnetism is the property of certain materials to attract iron, steel and other substances. This property is there in the magnets, in some metals and minerals and can also be produced with electric currents. For a long time the understanding of magnetism and of terrestrial magnetism has grown on equal footsteps being the Earth's magnetic field a natural colossal laboratory for magnetism. During the years the development of a specific 'natural magnetic philosophy' has been formulated, with its maximum shine in the 17th century. To this new philosophy scholars such as Gilbert, Copernicus, Kircher, Cabeo and others have contributed and attributed to natural magnetic forces properties sometimes well beyond their real capability.

The history of the evolution of our understanding of magnetism and Earth's magnetism is very long and articulated; few other scientific disciplines can probably compete with the abundance of references, applications and interpretations that can be found in magnetism. The search of the historical sources where mentions of magnetic properties started, brings us to very distant times and spaces. We will run here shortly through this long and exciting history.

9.1
The Very Ancient Times

Along the route that Marco Polo opened to the far eastern world, travelers, dealers, and from the second half of the 16th century, Christian missionaries journeyed. Among them the Jesuits, always very vigilant to scientific developments, were able to consult, to translate and to copy, some of the most important scientific works preserved in the Chinese library in Peking. Among the important findings they showed that the Chinese had discovered the existence of magnets, and the elementary phenomena of magnetism, probably already well over two thousand years before Christ. According to some documents the Chinese also knew about the properties of orientation of magnets in the Earth's magnetic field.

In the Chinese mythology, indeed, occurs the story of a battle in which Huang Ti, in the distant 2634 B.C., used a kind of mobile device able to point out the geographic south. With this magic chariot (Fig. 9.1) the emperor succeeded, in difficult seeing conditions, to find the direction in which his enemies were escaping and was thus able to chase and capture them. Even if the chariot, object of this legend, called "See nan" could recall a peculiar form of magnetic compass, it is not certain that its mechanism of operation were based on magnetic properties, and therefore that it was really a precursor of a modern compass. It was possibly a purely mechanical device, having therefore nothing to do with magne-

Fig. 9.1. Artist's reproduction of magic chariot called *See nan* which in an early Chinese legend was used by the emperor Huang Ti as a mobile device able to point south (from A. Schuck, *Der Kompass*, 1911)

tism. On various occasions a reconstruction of this device was attempted but the way in which the chariot could work was never clearly explained. Therefore, it is opinion of many authors, that this account, even if curious and interesting for its possible connections with the birth of compass, is most likely only a legend. Notwithstanding this murky episode however it seems that the earliest Chinese discovered the directional properties of magnetite at a very early date. The use that they made of this discovery was probably never wide spread, being those devices probably only devoted to the emperors and to the court. The first detailed description of a compass in China is found in a book called, the Mung khi Py Than, written in A.D. 1088, by Shon Kua. In another work entitled Pen-Thsao-yan, composed between 1111 and 1117, Khou Tsung-Shih reports on magnetic declination; the author observed that the compass does not perfectly lines up with the N-S direction, but attributes this observable fact to an imperfect support or suspension of the magnetic needle.

In the western world the term 'magnetism' takes its name from Magnesia, an ancient Greek city in Anatolia (in modern Turkey), where small parts of the mineral we now call magnetite could be found amongst other more common rocks. The Greeks and Romans, however, although knowing about the properties of magnetite, such as its ability to attract small pieces of iron (Plato makes a precise statement about it), never reported about anything that we would now can call polarity nor about the possibility of using the compass to give direction in the Earth's magnetic field. At sea, the

Greeks, Romans and other Mediterranean sailors, used only the Sun and the stars for orientation in navigation.

9.2
A Light in the Middle Ages

The first European written reference to magnetic properties and their application in the art of navigation is found in two essays, approximately written in 1190, "*De Utensilibus*" and "*De Naturis Rerum*" by the English monk Alexander Neckam. In the first part of the two mentioned works, the author describes the use of a magnetic needle to point to the north, and the use that can be made when sailing in the absence of the Sun or when the sky is cloudy. The second essay contains a description of what we can consider one of the first compasses, constituted by a rotating needle on a pivot. Neckam describes such objects as if they were not devices of common use; we can therefore infer that, at the time in Europe the use of the compass among sailors was still little known.

Magnetism knowledge was spreading in the reviving European cultural atmosphere of the time. Some mentions to this peculiar phenomenon can be found in very early literary pieces; for example in the first Italian language poems by Guido Guinizelli and even by Dante Alighieri, the spiritual writer. In the Canto XII of Paradise, the third section of *Divine Comedy* published in 1321, Dante was in conversation with a number of saints and when his attention was suddenly attracted to St. Bonaventura, he compares the turning of his own head, to listen to St. Bonaventura words, to the attraction of a magnet towards the north star

del cor de l'una de le luci nove
si mosse voce, che l'ago a la stella
parer mi fece in volgermi al suo dove

from the heart of one amongst the new lights
mov'd a voice, that made me seem like the needle
to the star, in turning to its whereabouts

The Englishman Roger Bacon, a Franciscan monk and his French contemporary, Petrus Peregrinus, were instead dwelling in the natural description of magnetic properties of matter, in a very efficient way: by 'experimenting'. Peregrinus's findings on magnetism are collected in a letter called *Epistola ad Sygerum de Foucaucourt militem, de magnete*, written in August 1269, when the author, fighting under Charles of Anjou, King of Naples and Sicily, was in a truce of a battle in southern Italy. The essay of Peregrinus is divided in two parts: in the first one the author describes that the determination of the two poles of opposite sign is possible on a sphere of magnetite, defining therefore for the first time, certainly in Europe, the concept of magnetic polarity. Moreover he describes how the opposite poles of a magnet are attracted each other, how a magnet receives its natural virtues, and he shows that the two magnetic polarities in a magnet cannot be separated. In the second part of its work Peregrinus describes two types of compass, that hanging to a mobile pivot and that in which the needle floats; he explains that a natural magnet can magnetize a magnetic needle and how to realize a tool to determine the azimuths of the Sun and the Moon and the shrewdness to improve it.

For what concerns the origin of the magnetic actions on the needles Peregrinus, believed that magnetic needles directly receive their virtue from the celestial poles;

Francis Bacon on the other hand, in his *Opus Minus* (1266), had already stated that magnets receive their virtue directly from the Earth, although not yet asserting that the Earth itself is a magnet. The Petrus Peregrinus Epistola is a document of particular importance since it is reasonably the first true scientific investigation in magnetism, if not the first scientific work in any discipline; it was circulated in 'manuscript' form from 1269 and then printed in 1588 (Fig. 9.2).

9.3
The Discovery of Declination and Inclination

Establishing the priority for the discovery of magnetic declination, as well as the invention of the compass itself, are both difficult. For declination several hypotheses and legends exist; one of the most credited, and probably the most realistic, sustains that Christopher Columbus, during his first voyage across the Atlantic, was the first to recognize both the existence of a declination angle as well as its variation in space. The observations made by Columbus in his first voyage in 1492, and reported in his journals, are consistent with his having crossed the agonic line (zero declination). However it is not known if he identifies this as an evidence of a general space-variation of magnetic declination. It is certain that in the first half of the 16th century the spatial variation of declination was well known and there were many explorers who measured it around the world, primarily at sea or in harbors. The exceptions to such a rule are not many and one of these is the well known measurement performed by Hartmann, a Jesuit, in Rome in 1510, reporting a value of 6° E. Hartmann was also the first that made account that the direction pointed out by the magnetic needle at rest is not parallel to the Earth's surface, introducing probably for the first time the concept of magnetic inclination. However the first explicit accurate measurement of inclination was made only several years later, in London in about 1576, by Robert Norman, a compass maker. The value for London was 71°50' and the magnetic needle tilted downwards its north pole. Norman's results were published in 1581 in a treatise 'The Newe Attractive', a work that today we could classify as popular science, since it was written in English for all people. The author, an autodidactic, published his essay hoping to allow people to take advantage from his discovery. To give an idea of the epoch feelings, Norman understanding the trouble that could come from natural truth investigations and in view of the consequences when the results were not in line with the approved dogmas, dedicated his essay to a notable, William Borough, superintendent of the Royal Navy, almost excusing himself fot having revealed obscure phenomena implying however that it was all done to the glory of God, and of course, also to enrich his native country.

The discovery of inclination was fundamental since it opened the road to the correct drawing of the Earth's magnetic field lines of force, and therefore to a more complete description of the field. Gerhard Kremer (Geraldus Mercator), Flemish mathematician and geographer was the first to have conceived the cartographic representation with meridians and parallels at right angles. He understood that if all magnetic lines, along which the magnetic needles were lined up, were extended over the surface of the Earth, they would meet in a single point that he called magnetic pole, distinguished from the geographic pole. To this place on Earth, Mercator attributed the source of the attractive force to which all magnetic needles are subject. Understanding of terrestrial magnetism phenomena began to diffuse in the western world at late Renais-

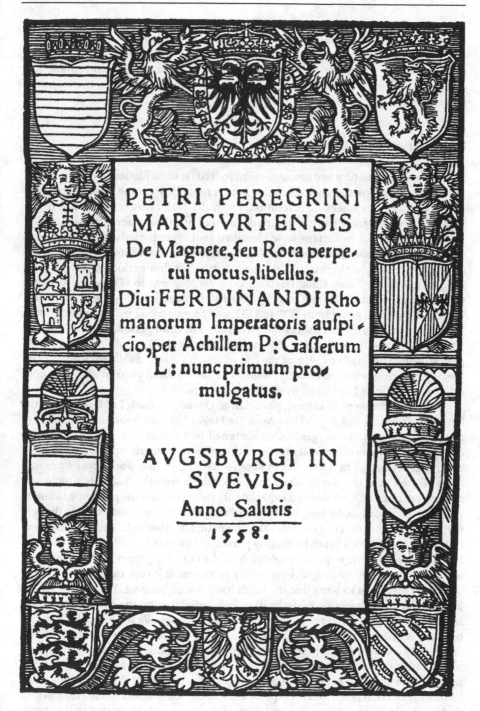

Fig. 9.2. Reproduction of front cover of the 1558 printed version of Petrus Peregrinus' *Epistola* written in August 1269 in which Peregrinus' findings on magnetism are collected

sance times. Natural philosophers and also occultists saw in magnetism the evident demonstration of the existence of mysterious and also spiritual powers. According to many philosophers, *aimant,* the French term still today in use to mean magnet (present participle of the verb *aimer,* to love), is a reminder of the mysterious and almost animal origin that was attributed to the phenomenon of attraction between two magnetic poles of opposite signs.

9.4
Geomagnetism in Gilbert's Epoch

As said above, magnetism and geomagnetism walked on same footsteps for a long time, the birth of geomagnetism as a discipline has its roots in the troubled developments of the 17th century scientific revolution. From the 17th century to the end of the 19th century, several scientists and philosophers put their efforts in its development. Some science historians have identified four main periods of this evolution, each one distinguished from the others by the presence of some important milestones on the nature and on the origin of geomagnetism. The first period goes from 1600, the year in which William Gilbert published his *De Magnete,* to 1777, when Augustin Coulomb gave a famous lecture on magnetism at the Académie des Sciences of Paris. The second period, going from 1777 to 1833, was characterized by the publication of Gauss's works, among them the famous *Intensitas.* The third period, going from about 1833 to 1873, saw the publication of Maxwell's treatise (1873). And finally the fourth period, going from 1873 to 1919, at the end of which Larmor published his paper on the origin of magnetism in rotating bodies. The last hundred years or so can be considered the period of the tangible, modern geomagnetism.

The impressive work of Gilbert, physician to Queen Elizabeth I of England, by the long title *Tractatus sive Physiologia Nova De Magnete, Magneticisque Corporibus et de Magno Magnete Tellure,* generally shortened to *De Magnete,* has been considered by many as the acme of centuries of thoughts and experimentations in magnetism, in general, and in particular in geomagnetism. In this work results are well exposed, and also the way in which the author deals with the matter is simple and comprehensible. Gilbert devised experiments for explaining the phenomena and to eliminate some diffused wrong beliefs about magnetism. It was for example thought that garlic had the power to influence the magnetic direction indicated by a compass. Gilbert disproved such belief by making personally an experiment on the topic! Gilbert also denied other popular beliefs such as the curative properties that magnets would have had, or the magnets capability to reconcile lovers, etc.

Gilbert asserts in its work that the Earth itself is a big magnet. The author defines and studies the distribution of declination and inclination on the Earth's surface using, as an analogy, a sphere of magnetite that he called *terrella* (little Earth). The recognition of magnetic inclination was considered particularly important by Gilbert since it was considered as a proof against the supporters, still numerous to the epoch, of a possible celestial magnetic influence. The positive polarity of magnetic needles in fact, in the northern hemisphere, bends toward the interior of the Earth rather than toward the celestial pole. Gilbert's experimental approach also brought him to discover that fire could destroy magnetic properties, when magnets were heated to the incandescence point. Gilbert also gave a cosmic interpretation of magnetism. He believed that the Earth rotated around a suitable axis, since he considered unreasonable to think that all the known stars of the 'primum mobile' were to rotate around the Earth. The alignment of the magnetic poles on the Earth forms a natural axis in space. A cosmic force, of magnetic nature, was therefore what maintains the Earth in rotation. All magnets receive their magnetization from the Earth, as the Earth received its own from outer space. Obviously Gilbert also had an answer to the enigma set by the mechanism that holds the solar system together: what held the planets and impede them from falling into the Sun or being lost in space, was magnetism. Probably this is all too much for magnetism to explain but certainly focussed a lot of attention on this phenomenon.

9.5
Secular Variation

The secular variation of the Earth's magnetic field is not mentioned in Gilbert's *De Magnete* since it was discovered only some years later. William Borough had conducted a study on the declination in London undertaking a series of measurements between 1580 and 1581; Edmund Gunter repeated the measurements in the same location in 1622 and found a different declination value, but it was only Henry Gellibrand who, through the comparison of these measurements with some of his own made in 1633 and 1634, realized that the difference among the data taken at different epochs, had to be attributed to a real physical phenomenon rather than to accidental errors of measurement. Later on in 1722, the characteristics of secular variation were carefully studied by George Graham, a clockmaker, through an accurate series of measurements undertaken with a special compass realized by himself.

At the end of the 17th century, therefore it was clear that the compass alone was not a suitable tool for navigation, especially on the great distances that, from this time on, concerned the whole terrestrial globe. To prevent this problem, magnetic data were collected from sailors in the different areas of the world. An example of such collection is Kircher's work of 1641, *Magnes sive de arte Magnetica*. To have a representation of spatial variations of declination, that were able to be a practical consultation tool for the sailors, magnetic cartography was introduced. The first magnetic maps consisted of a collection of declination values simply written on the geographic map, but only a few of them are available today. An example is the map produced in about 1635 by the Italian Cristoforo Borri in his work *De arte navigandi*, based on about twenty measurements. The paper is lost, but is mentioned by Kircher in his work.

The lack of magnetic measurements and the need of a practical consultation tool, were among the motivations that persuaded Edmund Halley to undertake two long journeys of declination observations in the Atlantic Ocean in 1698 and 1700. Halley was a well-known astronomer (everyone knows about the discovery of the comet named after him), but he was also interested in the study of both the spatial and time variations of Earth's magnetic field elements. Surely the most original result of his work was the production of the first map of declination for the Atlantic (Fig. 9.3) and later for almost the whole globe; these maps were realized with the introduction of equal value lines joining points having the same value of declination (now called isogonic

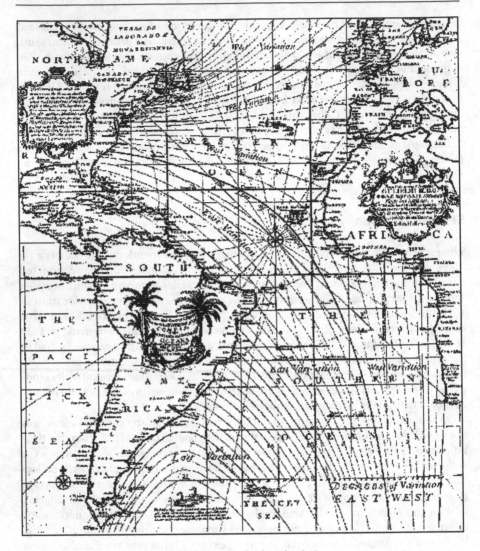

Fig. 9.3. Reproduction of Halley's declination map for the Atlantic Ocean

lines) reported directly on the geographic map. Because of the existence of secular variation, Halley also faced the need of a different interpretation of the cause of the Earth's magnetic field. His new theory foresaw the existence of two concentric nuclei inside the Earth, separated by a liquid layer in order to allow them to have an independent rotation speed. This theory explained the distribution of declination in the various places of the Earth, that he attributed to the distribution of magnetic masses, and the westward drift, as due to differential rotation.

The first map of inclination was made for the southwest of England in 1721 by William Whiston. In 1768, Wilcke realized an inclination chart for the whole globe. The secular variation showed a systematic westward drift of all the isolines of declination. In 1858 finally, the British Admiralty started to produce a series of magnetic maps of the globe, updated at regular time intervals.

9.6
Geomagnetism from Gauss Onwards

Probably the most important events in the history of physics directly connected to Earth's magnetism are those that took place at the beginning of the 19th century. Between 1832 and 1840, in fact, Carl Friedrich Gauss published three memoirs in which he developed a theory of geomagnetism and for the first time gave a procedure for the measurement of magnetic field intensity, using only fundamental mechanical units. It was the first time in the history of physics that a non-mechanical quantity was expressed in terms of mechanical ones: to this end, Gauss designed a new instrument, the 'magnetometer'. He also gave the general theory according to which forces vary according to the inverse square of distance, and understood that magnetic dipole forces vary according to the inverse cube of distance.

To check the validity of his new geomagnetic theory, Gauss started an international cooperative project, in which about seventy institutions spread all over the world took part. This was the first international cooperation in geophysics. In Gauss's international cooperative project, called *Magnetischer Verein* (Magnetic Union), aimed to measure the Earth's magnetic field simultaneously at many locations, the requirement was that measurements were carried out on certain days of the year, according to specified procedures and with particular instruments. From 1838 onwards, the cooperation improved, thanks to the *Magnetic Crusade* organized by the British Association for the Advancement of Science and the Royal Society of London.

Gauss's work *Allgemeine Theorie des Erdmagnetismus* (General Theory of the Earth's Magnetism) represents the theoretical counterpart of the collection of geophysical data. It marked the beginning of modern mathematical analysis of the magnetic field at the Earth's surface. In this memoir, Gauss supposed that the geomagnetic force is derivable from a potential, expressed as the sum of spherical harmonics. Before him, other scientists, Legendre in particular, had used spherical harmonic analysis, but Gauss's application of the theory concerned data that covered a long period of time and were made with several instruments and different procedures. Following this success it can reasonably be stated that geomagnetism became an independent discipline, stemming from both, physics and natural philosophy.

While Gauss's analysis had the aim of improving the accuracy of the instantaneous, global description of the geomagnetic field and of its spatial and temporal variation, through the collection and the analysis of geomagnetic data, the electromagnetic theories inspired by Maxwell focused attention on the behavior of the field and its causes. Maxwell's *Treatise on Electricity and Magnetism* provided a new mathematical instrument for the analysis of electrical and magnetic phenomena and a physical model based on the theory of the field. After Maxwell, the expression "magnetic force" turned into "magnetic field" in the geomagnetic scientists's lexicon. In the last two decades of the 19th century, detailed mathematical description using Gauss's analysis and Maxwell's electromagnetic theory, constituted the shared scientific values around which the geomagnetic community was established: geomagnetic researches increased greatly both in data collection and in the elaboration of physical models. These events mark the birth of a new generation of geomagnetic scientists. A new journal, *Terrestrial Magnetism*, was founded in 1896, to publish the increasing number of articles on the subject. This journal has now become the well known *Journal of Geophysical Research* that includes all the geophysical sciences.

The last hundred years or so of the history of geomagnetism are strictly tied to a variety of observations and 'neighboring' scientific developments. Solar research, for example, has brought observations that were crucial for the understanding of magnetic storms, polar aurora, diurnal variation and many other time variations of the Earth's magnetic field. The discovery of the 11-year sunspot cycle and the introduction of the empirical Zurich sunspot number, have given birth to a long series of studies that finally allowed to correctly connect geomagnetic phenomena and solar science. The correct interpretation that sunspots are connected to solar magnetism suggested to Larmor that a dynamo mechanism was acting in the interior of the Sun. The same mechanism could work for the Earth, if only it possessed an internal electrically conducting fluid core. The discovery that this is the case came some years later from seismology, a sister science, noting that seismic transverse waves were not propagating in the Earth's outer core, and with the help of geochemists that believed that in Earth's history iron had sunk towards the center.

In the second half of the 20th century, space physics was a major actor on the scene. Investigations by means of space probes, at first around the Earth and later in the whole solar system, have enriched the limited ground magnetic measurements with a variety of new observations that have significantly enlarged our views. The discovery of the magnetosphere and the interaction with solar wind, the discovery of the Van Allen belts and the relation of space plasma dynamics to Earth's magnetic field variations, have been possible only with these unique *in situ* observations. For example, no one was expecting a long magnetic tail on the Earth's night side, a tail into which frequently the Moon, in its orbit around the Earth, enters. Probably the most striking discovery of the recent space age, for what concerns geomagnetism, is the awareness that other solar system objects, including planets, possess a magnetic field. Planetary bodies, also very different from the Earth, have their own magnetic fields. Planets like Mercury, Jupiter, Saturn, Uranus and Neptune have a significant measurable magnetic field, others may have had such fields in their past (Mars, the Earth's Moon). At least in our solar system, planetary magnetism seems to be the rule rather than the exception. Some scientists illustrate in their papers this phenomenon in a kind of comparative magnetic planetology.

9.7
Rock Magnetism

In the 19th century, observations on volcanic rocks and baked clays magnetism, started in and continuing sporadically since the 17th century, became progressively more systematic and in the early 20th century, thanks to work by Folgerhaiter, Brunhes, Mercanton, Matuyama, Königsberger and many others, it was an acquired fact that these materials record the direction of the Earth's magnetic field when they cool down. Two clearly established results were the capability of dating earthenware artefacts according to secular variation and the fact that the direction of magnetization of many

Pleistocene volcanic rocks was opposite with respect to the polarity of the present-day magnetic field. However, while Thellier developed archaeomagnetic techniques in France, rock magnetism was still something exotic, with few, if any, applications in the geological field. On the one hand, the low sensitivity of instruments allowed reliable measurements only on volcanic rocks. On the other hand, basic knowledge about rock magnetism was still insufficient.

Paleomagnetism started in the fifties of the 20th century, thanks to a renewed theoretical interest and new instrumental capabilities. One of these came from the highly sensitive astatic magnetometer developed by Blackett to verify the hypothesis that a magnetic field is a fundamental property of any rotating body. The outcome of the experiment (1952) was negative, but the instrument was there and could still be used, this time to measure the magnetization of rocks, even at low intensities. More or less at the same time, some physicists, among them Néel, Nagata and Stacey, were laying renewed foundations for a physical theory of remanent magnetization that would support experimental observations.

Observations on sedimentary rocks, coupled with the ability to date volcanic rocks using isotopic methods, opened up broad horizons. Thanks to work by Blackett's and Runcorn's groups in the late fifties in England, it was already clear that paleomagnetic data spoke in favor of large displacements of continental masses over geological times. The first polarity inversion time scales were produced in the early sixties, with the contribution of very many scholars: among them, we recall Doell, Cox, Watkins, McDougall, Tarling, Wensink. The first systematic applications of magnetostratigraphy were due to Khramov, in Russia, while the study of deep-sea cores was mainly developed by Opdyke, at the Lamont-Doherty Laboratory.

In the early sixties, paleomagnetic data collected from various continents were already sufficient to demonstrate that the opening of the Atlantic Ocean and the Gondwana continent were not merely some nice working assumptions, but a set of phenomena, documented by precise scientific observations. This evidence was not accepted immediately: after years spent first to demolish Wegener's theory, then to ignore it as something best not mentioned in geology's elegant circles, something more was required to get minds moving. The impetus came once again from magnetism, this time from the ocean floor. The Vine-Matthews and Morley model (1963) allowed to quantify the sea-floor expansion hypothesis and became one of the keystones of plate tectonics theory.

Paleomagnetism was welcomed among the geological sciences thanks to its essential contribution to global dynamics, but another enormously important result should not be forgotten. The joint development of measurements of oceanic anomalies and magnetic stratigraphy, both on oceanic piston cores and on land sections, produced Geomagnetic Polarity Time Scales (GPTS), something that today seems commonplace and is used by everyone, but in fact it is the first great synthesis of an absolute chronology for the last 200 Myr in the history of the Earth.

The final decades of the twentieth century saw paleomagnetism rapidly developing along more differentiated directions. First-generation paleomagnetists were mostly interested in the time aspects, to decipher the history of the magnetic field written in the rocks. The second generation was that of rock-magnetists, who gave priority to thoroughly studying the magnetic properties of rocks: the processes whereby they acquire a remanent magnetization, the characteristics of minerals, the magnetic fabric and so on. The current, third generation can be called that of magnetists without any hyphens, because it is expanding its interests into all fields where magnetic measurement techniques can be applied: climate, environment, biomagnetism, etc. Today, it is not uncommon for a "paleomagnetic" laboratory to carry out precious little "paleo" work. One of the most fascinating aspects of this very short history is that paleomagnetism, which originated as a super-specialist discipline, has now become a typical interdisciplinary field of studies.

Suggested Readings and Sources of Figures

Books

Jonkers ART (2003) Earth's magnetism in the age of sail. Johns Hopkins University Press, Baltimore, 300 pp
Merrill RT, McElhinny MW, McFadden PL (1996) The magnetic field of the Earth: Paleomagnetism, the core and the deep mantle. Academic Press, San Diego, California, 531 pp

Articles

Hellmann G (1909) Magnetische Kartographie in historisch kritischer Darstellung. Abh. Kön. Preussischen met. Inst. 3(No. 3), 61 pp
Mitchell A, Crichton (1937) Chapters in the history of terrestrial magnetism. Chapter II – The discovery of the magnetic declination. Terr Magn Atmos Electr 42:241–280
Needham J (1962) Science and civilisation in China. Vol 4: Physics and physical technology. Part I: Physics. Cambridge University Press, Cambridge pp 229–234
Stern DP (2002) A millennium of geomagnetism. Rev Geophys 40:3

Appendix

Magnetic Quantities, SI and cgs Units, Conversion Factors

Table A.1. Magnetic quantities, SI and cgs units and conversion factors

Quantity	Symbol	SI unit	cgs unit	Conversion factor
Magnetic induction	B	Tesla: T	Gauss: G	$1\,T = 10^4\,G$
Magnetic field strength	H	$A\,m^{-1}$	Oersted: Oe	$1\,A\,m^{-1} = 4\pi\,10^{-3}\,Oe$
Magnetic moment	M, m	$A\,m^2$	emu: $G\,cm^3$	$1\,A\,m^2 = 10^3\,emu$
Permeability of free space	μ_0	Henry m^{-1}: $H\,m^{-1}$	dimensionless	$1\,H\,m^{-1} = 10^7/(4\pi)\,cgs$
Intensity of magnetization (= magnetic moment per volume)	J	$A\,m^{-1}$	emu cm^{-3}: G	$1\,A\,m^{-1} = 10^{-3}\,emu\,cm^{-3}$
Magnetic susceptibility, by volume	κ	dimensionless	dimensionless	$1\,SI = 1/(4\pi)\,cgs$
Magnetic susceptibility, by mass	χ	$m^3\,kg^{-1}$	$cm^3\,g^{-1}$	$1\,m^3kg^{-1} = 10^3/(4\pi)cm^3g^{-1}$

Magnetic Quantities, SI and cgs Units, Conversion Factors

Index